Spray Simulation

Spray forming combines the metallurgical processes of metal casting and powder metallurgy to fabricate metal products with enhanced properties. This book provides an instruction to the various modelling and simulation techniques employed in spray forming, and shows how they are applied in process analysis and development.

The author begins by deriving and describing the main models. He then presents their application in the simulation of the key features of spray forming. Wherever possible he discusses theoretical results with reference to experimental data. Building on the features of metal spray forming, he also derives common characteristic modelling features that may be useful in the simulation of related spray processes.

The book is aimed at researchers and engineers working in process technology, chemical engineering and materials science.

Udo Fritsching received his Ph.D. from the University of Bremen, Germany, and is currently head of the Research Group at the Institute for Materials Science and apl. Professor at the University of Bremen. He is the author or coauthor of 160 scientific papers and has five patent applications pending.

Spray Simulation

Modelling and Numerical Simulation of Sprayforming Metals

Udo Fritsching

Universität Bremen

CAMBRIDGE UNIVERSITY PRESS
Cambridge, New York, Melbourne, Madrid, Cape Town, Singapore, São Paulo

Cambridge University Press
The Edinburgh Building, Cambridge CB2 8RU, UK

Published in the United States of America by Cambridge University Press, New York

www.cambridge.org
Information on this title: www.cambridge.org/9780521820981

First published 2004
This digitally printed version 2007

A catalogue record for this publication is available from the British Library

Library of Congress Cataloguing in Publication data
Fritsching, Udo, 1959–
Spray simulation: modelling and numerical simulation of sprayforming metals / Udo Fritsching.
    p. cm.
ISBN 0 521 82098 7
1. Metal spraying – Mathematical models.   2. Metal spraying – Computer simulation.   I. Title.
TS655.F75   2003
6.71.7′34 – dc21   2003055191

ISBN 978-0-521-82098-1 hardback
ISBN 978-0-521-03777-8 paperback

Contents

Preface   *page* vii
Nomenclature   x

1 Introduction   **1**

2 Spray forming of metals   **6**

   2.1   The spray forming process   6
   2.2   Division of spray forming into subprocesses 10

3 Modelling within chemical and process technologies   **21**

4 Fluid disintegration   **26**

   4.1   Melt flow in tundish and nozzle   28
   4.2   The gas flow field near the nozzle   43
   4.3   Jet disintegration   67

5 Spray   **94**

   5.1   Particle movement and cooling   97
   5.2   Internal spray flow field   121
   5.3   Spray-chamber flow   144
   5.4   Droplet and particle collisions   147

6   Compaction   161

6.1   Droplet impact and compaction   161
6.2   Geometric modelling   176
6.3   Billet cooling   187
6.4   Material properties   218

7   An integral modelling approach   233

8   Summary and outlook   243

Bibliography   245
Useful web pages   269
Index   271

Preface

This book describes the fundamentals and potentials of modelling and simulation of complex engineering processes, based on, as an example, simulation of the spray forming process of metals. The spray forming process, in this context, is a typical example of a complex technical spray process. Spray forming, basically, is a metallurgical process whereby near-net shaped preforms with outstanding material properties may be produced direct from a metal melt via atomization and consolidation of droplets. For proper analysis of this process, first successive physical submodels are derived and are then implemented into an integrated coupled process model. The theoretical effects predicted by each submodel are then discussed and are compared to experimental findings, where available, and are summarized under the heading 'spray simulation'. The book should give engineering students and practising engineers in industry and universities a detailed introduction to this rapidly growing area of research and development.

In order to develop an integral model for such technically complex processes as the spray forming of metals, it is essential that the model is broken down into a number of smaller steps. For spray forming, the key subprocesses are:

- atomization of the metal melt,
- dispersed multiphase flow in the spray,
- compaction of the spray and formation of the deposit.

These subprocesses may be further divided until a sequential (or parallel) series of unit operational tasks is derived. For these tasks, individual balances of momentum, heat and mass are to be performed to derive a fundamental model for each. In addition, some additional submodels need to be derived or applied. The general description of this modelling approach to the spray forming process is the fundamental aim of this book, which therefore:

- introduces a general modelling and simulation strategy for complex spray processes,
- reviews relevant technical contributions on spray form modelling and simulation, and
- analyses and discusses the physical behaviour of each subprocesses and materials in the spray forming process.

This work is based on a number of investigations of spray forming carried out by researchers all over the world. Major contributions have been given from research projects

conducted by the author's research group on 'multiphase flow, heat and mass transfer' at the University of Bremen, the Foundation Institute for Material Science (IWT), as well as the Special Research Cooperation Project on spray forming SFB 372 at the University of Bremen. These projects have been funded, for example, by the Deutsche Forschungsgemeinschaft DFG, whose support is gratefully acknowledged. Several graduate and PhD students contributed to this project. I would like to thank all of them for their valuable contributions, especially Dr.-Ing. O. Ahrens, Dr.-Ing. D. Bergmann, Dr.-Ing. I. Gillandt, Dr.-Ing. U. Heck, Dipl.-Ing. M. Krauss, Dipl.-Ing. S. Markus, Dipl.-Ing. O. Meyer and Dr.-Ing. H. Zhang. Also, I would like to thank those guests whom I had the pleasure of hosting at the University of Bremen and who contributed to the development of this book, namely Professor Dr.-Ing. C. T. Crowe and Professor Dr. C. Cui. I acknowledge Professor Dr.-Ing.

K. Bauckhage for initiating research in this field and thank him for his continuous support of research in spray forming at the University of Bremen.

I would like to thank my family, Karin and Anna, for their understanding and support.

In order to keep the price of this book affordable, it has been decided to reproduce all figures in black/white. All coloured plots and pictures can be found and downloaded by interested readers from the author's homepage. Some of the spray simulation programs used in this book may also be downloaded from this web page. The URL is:
www.iwt-bremen.de/vt/MPS/

Nomenclature

A	area	m^2
a_i	coefficients	
a	temperature conductivity	m^2/s
b_i	coefficients	
c_d	resistance (drag) coefficient	
c_p	specific heat capacity	kJ/kg K
c_1, c_2, c, c_T	constants of turbulence model	
D_k	dissipation	
d, D	diameter	m
$d_{3.2}$	Sauter mean diameter, SMD	m
d_{max}	maximum spread diameter	m
Ec	Eckart number	
F	force	N
F_f	volume ratio, filling function	
f_r	coefficient of friction, normalized resistance	
$f_{s,l}$	solid or liquid content	
f	frequency	1/s
f	distribution density of particles	
G	coefficient for interparticulate forces	
G	number of solid fragments	
g	gravity constant	m/s^2
$\dot{g}$	growth rate	m/s
GMR	mass flow rate ratio gas/metal	
H, h	height	m
H	enthalpy	kJ
h	specific enthalpy	kJ/kg
h_f	film thickness	m
h_l	ligament height	m
I, K	Bessel function	
J	nucleation rate	1/s
k_S	empirical constant	
k	turbulent kinetic energy	m^2/s^2
k_p	compaction rate	

k_B	Boltzmann constant	J/K
L	length	m
L_h	latent heat of fusion	kJ/kg
L_T	dissipation length scale	m
La	Laplace number	
l	length, distance to nozzle	m
M	fragmentation number	
$\dot{M}$	mass flow rate	kg/s
Ma	Mach number	
m	mass	kg
m	mode	
$\dot{m}$	mass flux	kg/m^2 s
Nu	Nusselt number	
N, n	number concentration, particle number	1/m^3
Oh	Ohnesorge number	
P	number of collisions	
Pe	Peclet number	
p	pressure	Pa
p	microporosity function	
q_r	probability density function	1/m
$\dot{Q}$	heat flow rate	W
$\dot{q}$	heat flux	W/m^2
r	radial coordinate	m
$r_{0.5}$	half-width radius	m
R	gas constant	kJ/kg K
R_L	Lagrangian time correlation coefficient	
Re	Reynolds number	
Real	real part	
S	source/sink	
Sha	Shannon entropy	
St	Stokes number	
Ste	Stefan number	
s	path	m
T	temperature	K
T^*	Stefan number	
$\dot{T}$	cooling rate (velocity)	K/s
ΔT	temperature difference	K
ΔT	undercooling	K
t	time	s
u, v, w	velocity components	m/s
V	volume	m^3
v	velocity of solidification front	m/s

$\mathbf{W}, \mathbf{F}, \mathbf{G}, \mathbf{Q}$	matrix	
We	Weber number	
x, y, z	plane Cartesian coordinates	m
x_K	length of supersonic core	m
x_s	mean distance between solid fragments	m
Z^*	splashing number	
z	distance atomizer – substrate	m
z, r, θ	cylindrical coordinates	m, m, °
α_G	gas nozzle inclination angle	°
α_f, α_g	volumetric content of gas, liquid	
α_{spray}	spray inclination angle	°
α	heat transfer coefficient	W/m^2 K
Γ	diffusivity	
Γ_S	Gamma function	
γ	solid–liquid surface tension	N/m
δ	excitation wavelength	m
δ	width of gas jets	m
ε	dissipation rate of turbulent kinetic energy	m^2/s^3
ε_S	radiation emissivity	
η_S	amplitude function of perturbation	
η_{ab}, η_B	amplitude of surface waves	
Θ_{col}	impact angle	°
θ	contact angle	°
θ	modified temperature	K
κ	isentropic exponent	
κ_0	surface curvature	1/m^2
λ	heat conductivity	W/m K
λ_0	reference heat conductivity	W/m K
λ_d	wavelength	m
λ_e	solidification coefficient	
μ	dynamic viscosity	kg/m s
ν	kinematic viscosity	m^2/s
ν_m	molar volume	m^3/mol
ξ_g	boundary layer coefficient	
ξ, η	dimensionless coordinates	
ρ	density	kg/m^3
σ_l	surface tension	N/m
σ_d	logarithmic standard deviation	
$\sigma_h, \sigma\frac{\varepsilon}{x}, \sigma_k$	constants of turbulence model	
σ_S	Stefan–Boltzmann constant	W/m^2 K^4
σ_t	relative turbulence intensity	
τ	shear stress	N/m^2
τ_p	relaxation time	s

τ_T	eddy lifetime	s
τ_u	passing time through eddy	s
τ_v	interaction time	s
Φ	transport variable	
Φ	velocity potential	
Φ	impact angle	°
φ	velocity number	
χ	impact parameter	m
Ψ	stream function	
ψ_f	function of fluid density	
Ω	collision function	
ω	growth rate	1/s

Indices

A	nozzle exit area
a	lift
a	outer side
abs	total value
b	Basset
c	centre-line
c, crit	critical
ct	contact layer
cyl	cylindrical
d	dispersed phase (droplet)
eff	effective value
ener	energy
f	film
f	fluid
g	gas phase
g	gravity
h	hydrostatic
het	heterogeneous
hom	homogeneous
i	imaginary part
i	inner side
i, j	numbering, grid index
ideal	ideal state
in	inflow
jacket	side region of billet
k	nucleation
k	compaction
Lub	Lubanska
l	liquid

l	liquidus
m	mass
m	mean value
min, max	minimum value, maximum value
mom	moment
n	normal direction
out	melt exit
por	porosity
p	particle
p	pressure
p	projected
r	real part
rel	relative value
s	solidus
s	spray
Sh	shadow
sin	sinus
S	melt
t	turbulence
t	inertia
t	tangential direction
top	top side of billet
tor	torus
u	environment
v	velocity
w	wall
w	resistance
zu	addition
0	stagnation value
1	primary gas
2	secondary (atomization) gas
*	critical condition

1 Introduction

Modelling of technical production facilities, plants and processes is an integral part of engineering and process technology development, planning and construction. The successful implementation of modelling tools is strongly related to one's understanding of the physical processes involved. Most important in the context of chemical and process technologies are momentum, heat and mass transfer during production. Projection, or scaling, of the unit operations of a complex production plant or process, from laboratory-scale or pilot-plant-scale to production-scale, based on operational models (in connection with well-known scaling-up problems) as well as abstract planning models, is a traditional but important development tool in process technology and chemical engineering. In a proper modelling approach, important features and the complex coupled behaviour of engineering processes and plants may be simulated from process and safety aspects viewpoints, as well as from economic and ecologic aspects. Model applications, in addition, allow subdivision of complex processes into single steps and enable definition of their interfaces, as well as sequential investigation of the interaction between these processes in a complex plant. From here, realization conditions and optimization potentials of a complex process or facility may be evaluated and tested. These days, in addition to classical modelling methods, increased input from mathematical models and numerical simulations based on computer tools and programs is to be found in engineering practice. The increasing importance of these techniques is reflected by their incorporation into educational programmes at universities within mechanical and chemical engineering courses.

The importance of numerical models and simulation tools is increasing dramatically. The underlying physical models are based on several input sources, ranging from empirical models to conservation equations for momentum, heat and mass transport in the form of partial differential or integro differential equations. Substantial development of modelling and simulation methods has been observed recently in academic research and development, as well as within industrial construction and optimization of processes and techniques. For the process or chemical industries, some recent examples of the successful inclusion of modelling and simulation practice in research and development may be found, for example, in Birtigh *et al.* (2000). This increasing importance of numerical simulation tools is directly related to three different developments, which are individually important, as is the interaction between them:

- First, the potential of numerical calculation tools has increased due to the exploding power of the computer hardware currently available. Not only have individual single processor computers increased their power by orders of magnitude in short time scales, but also interaction between multiple processors in parallel machines, computer clusters or vector computers has recently raised the hardware potential dramatically.

- Next, and equally important, the development of suitable sophisticated mathematical models for complex physical problems has grown tremendously. In the context of the processes to be described within this book, a variety of new complex mathematical models for the description of exchange and transport processes in single- and multiphase flows (based on experimental investigations or detailed simulations) has recently been derived.
- Last, but not least, developments in efficient numerical analysis tools and numeric mathematical methods for handling and solving the huge resulting system of equations have contributed to the increasing efficiency of simulated calculations.

Based on state-of-the-art modelling and simulation tools, a successful and realistic description of relevant technical and physical processes is possible. This story of success has increased the acceptance of numerical simulation tools in almost all technical disciplines. Closely connected to traditional and modern theoretical and experimental methods, numerical simulation has become a fundamental tool for the analysis and optimization of technical processes.

The process of spray forming, which will be discussed here in terms of modelling and simulation, is basically a metallurgical process, but will be mainly described from a fundamental process technology point of view. Metal spray forming and the production of metal powders by atomization, i.e. the technical processes evaluated in this book, are fundamentally related to the disintegration of a continuous molten metal stream into a dispersed system of droplets and particles. Atomization of melts and liquids is a classical process or chemical engineering operation, whereby a liquid continuum is transformed into a spray of dispersed droplets by intrinsic (e.g. potential) or extrinsic (e.g. kinetic) energy. The main purpose of technical atomization processes is the production of an increased liquid surface and phase boundary or interfacial area between liquid and gas. All transfer processes across phase boundaries directly depend on the exchange potential, which drives the process, and the size of the exchange surface. In a dispersed system, this gas/liquid contact area is equal to the total sum of surfaces of all individual drops, i.e. of all droplets within the spray. By increasing the relative size of the phase boundary in a dispersed system, the momentum, heat and mass transfer processes are intensified between the gas and the liquid. The total exchange flux within spray systems may thereby be increased by some orders of magnitude.

Atomization techniques in process technology or chemical engineering processes/plants can be applied to:

- impact-related processes, and
- spray-structure-related processes.

Some examples of spray process applications in engineering following this subdivision are listed below.

- Impact-related spray applications requiring a continuous fine spatial distribution of a liquid continuum, e.g. in the field of coating applications:

- for protection of metallic surfaces from corrosion in mechanical engineering through paint application;
- for coating of technical specimens/parts for use in private or industrial applications, including surface protection and colour (paint) application;
- for thermal plasma- or flame-spraying of particles to provide protective ceramic or metallic coatings in the metal industry;
- for spray granulation or coating of particles (for example in pelletizers or in fluidized beds), e.g. for pharmaceutical or food industry applications;
- for spray cooling in steel manufacturing or in spray heat treatment of metallic specimens.
- Spray-structure-related applications whereby the structure or properties of liquids or particulate solids are altered by gas/dispersed phase exchange processes, within:
  - thermal exchange processes, e.g. for rapid cooling and solidification of fluid metal melts in metal powder generation;
  - coupled mass and thermal transfer processes, e.g. spray drying (or spray crystallization) in food or dairy industries or in chemical mass products production;
  - particle separation from exhaust gases, e.g. from conventional power plants (wet scrubbing);
  - reaction processes within fuel applications in energy conversion, automotive, or aeroplane or aerospace engine or fuel jet applications.
- Combination of impact- and spray-structure-related spray process applications:
  - within droplet-based manufacturing technologies, e.g. for rapid prototyping;
  - for the generation of specimens and preforms by spray forming of metals.

In spray forming, a combination of nearly all the features, subprocesses and examples of atomization processes listed, may be found. Spray forming is, in its unique composition, an ideal and typical example of a complex technical atomization process. The numerical modelling and simulation techniques derived for analysis and description of the spray forming process may be easily transferred to other atomization and spray process applications.

In a first analysis approach, the complex coupled technical process is subdivided into single steps for further study. In the context of spray forming, subdivision of the technical atomization process into modular subprocesses can be done. This is illustrated in Figure 1.1, where the three main subprocesses discussed below are shown:

- atomization: the process of fluid disintegration or fragmentation, starting with the continuous delivery of the fluid or melt, and necessary supporting materials (such as gases or additives), to the resulting spray structure and droplet spectrum from the atomization process;
- spray: the establishing and spreading of the spray, to be described by a dispersed multiphase flow process with momentum, heat and mass transfer in all phases, and the exchange between the phases, as well as a possible secondary disintegration process of fluid ligaments or coalescence of droplets;

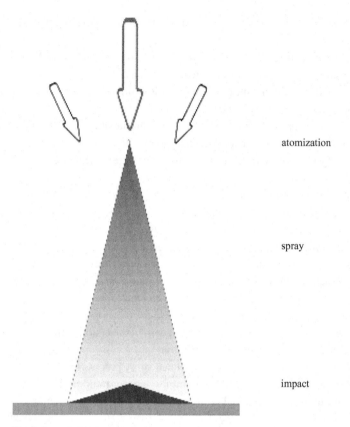

atomization

spray

impact

Fig. 1.1 Subdivision of an atomization process into subprocesses

- impact: the impact of spray droplets onto a solid or liquid surface and the compaction and growth of the impacting fluid or melt mass, as well as the building of the remaining layer or preform.

Central, integrative and common to all spray-related subprocesses is the fluids engineering and process or chemical engineering discipline of fluid dynamics in multiphase flows involving integral heat and mass transfer. The fundamental properties and applications of this discipline are central to the theme of this book.

Based on this method of analysis, modelling and numerical simulation are introduced as scientific tools for engineering process development, as applied to metal spray forming. Then, the individual physical processes that affect spray forming are introduced and implemented into an integral numerical model for spray forming as a whole. Recent modelling and simulation results for each subprocess involved during metal spray forming are discussed and summarized. Where possible, simulated results are compared to experimental results during spray forming, to promote physical understanding of the relevant subprocesses, and are discussed under the heading 'spray simulation'.

Numerical modelling and simulation of the individual steps involved in spray forming are presently of interest to several research groups in universities and industry worldwide. In this book, the current status of this rapidly expanding research area will be documented. But despite the emphasis given to metal spray forming processes, it is a major concern of this book to describe common analysis tools and to explain general principles that the reader may then apply to other spray modelling strategies and to other complex atomization and spray processes.

To enhance the general integral spray forming model further, additional physical sub-models need to be developed and boundary conditions determined. It is hoped that the combination of experimental, theoretical and numerical analyses presented here will contribute to the derivation and formulation of such additional subprocess models. Integration of these models into a general operational model of the spray forming process will then be possible.

In conclusion, the main aims of this book are:

- to introduce a general strategy for modelling and simulation of complex atomization and spray processes,
- to review relevant contributions on spray form modelling and simulation, and
- to analyse and discuss the physical behaviour of the spray forming process.

2 Spray forming of metals

In this chapter, fundamental features of the metal spray forming process are introduced in terms of their science and applications. Chapter 1 saw the division of the process into three main steps:

(1) disintegration (or atomization),
(2) spray establishment, and
(3) compaction.

Now, a more detailed introduction to those subprocesses that are especially important for application within the spray forming process, will be given.

2.1 The spray forming process

Spray forming is a metallurgical process that combines the main advantages of the two classical approaches to base manufacturing of sophisticated materials and preforms, i.e.:

- metal casting: involving high-volume production and near-net shape forming,
- powder metallurgy: involving near-net shape forming (at small volumes) to yield a homogeneous, fine-grained microstructure.

The spray forming process essentially combines atomization and spraying of a metal melt with the consolidation and compaction of the sprayed mass on a substrate. A typical technical plant sketch and systematic scheme of the spray forming process (as realized within several technical facilities and within the pilot-plant-scale facilities at the University of Bremen, which will be mainly referenced here) is illustrated in Figure 2.1. In the context of spray forming, a metallurgically prepared and premixed metal melt is distributed from the melting crucible via a tundish into the atomization area. Here, in most applications, inert gas jets with high kinetic energy impinge onto the metal stream and cause melt disintegration (twin-fluid atomization). In the resulting spray, the droplets are accelerated towards the substrate and thereby cool down and partly solidify due to intensive heat transfer to the cold atomization gas. The droplets and particles in the spray impinge onto the substrate thereby consolidating and depositing the desired product.

The basic concept of metal spray forming was established in the late 1960s in Swansea, Wales, by Singer (1970; 1972a,b) and coworkers. In the 1970s, the spray forming process was further developed as an alternative route for the production of thin preforms directly

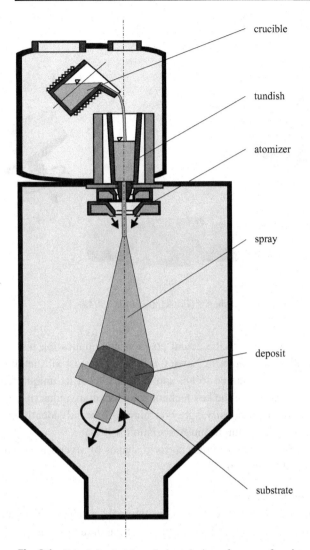

crucible

tundish

atomizer

spray

deposit

substrate

Fig. 2.1   Principle sketch and plant design of a spray forming process facility

from the melt. The spray forming process developed as a substitute for the conventional production processes of casting and subsequent hot and cold rolling of slabs (often in combination with an additional thermal energy source). The spray forming process was first used commercially by a number of Singer's young researchers (Leatham *et al.*, 1991; Leatham and Lawley, 1993; Leatham, 1999), who founded the company Osprey Metals in Neath, Wales. For this reason, the spray forming process is sometimes referred to as the Osprey process. Since then, worldwide interest in the physical basics and application potential of the spray forming process has spawned several research and development programmes at universities and within industries. An overview of the resulting industrial applications and the aims of industrial spray forming are given, for example, in Reichelt (1996) or Leatham

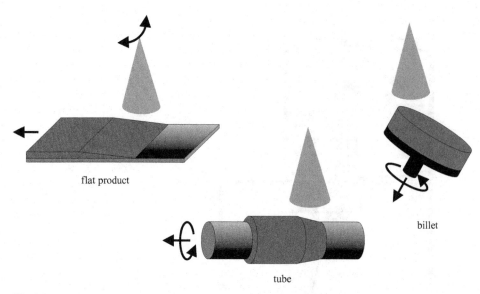

flat product

billet

tube

Fig. 2.2 Spray forming of different preform shapes (Fritsching *et al.*, 1994a)

(1999) and Leatham *et al.* (1991). Also, the actual position and application potential of spray forming are reviewed, for example, in Lawley (2000). In almost all metallurgical areas, spray forming has been, or is aimed to be, applied because of its unique features and potentials. It is referred to as one of the key technologies for future industrial applications. For example, in the aluminium industry, spray forming was recently identified as the 'highest priority' research area (The Aluminium Association, 1997).

In technical applications, nowadays, different preform shapes of several materials and alloys are produced via spray forming, such as:

- conventional metallic materials and alloys;
- materials and alloys that tend to segregate within conventional casting processes and are therefore complicated to handle, such as some alloys cast on an aluminium, copper or iron basis;
- super alloys (e.g. on a nickel base) for applications in aeroplane and aerospace industries;
- intermetallic composite materials (IMCs);
- metal-matrix-composite materials (MMCs), e.g. ceramic particle inclusions in a metal matrix.

The main geometries of spray formed preforms and materials produced at present are summarized in Figure 2.2. These include the following:

- Flat products that are formed by distributing the spray over a specific area of a flat substrate that is moved linearly. The spray is distributed either by oscillating (or scanning) the atomizer or is based on atomization of the melt within so-called linear nozzles, where the melt flow exits from the tundish via an elongated slit. Flat products are of interest for several technical applications, but, until now, problems associated with process reproducibility and control in flat product spray forming have meant that the industrial application of this type of spray forming is not always practicable.

- Tubes or rings are spray formed by applying the metal spray in a stationary or scanning movement onto a cylindrical or tube-shaped substrate. The inner material (base tube) may either be drilled off afterwards or, for realization of a twin-layered clad product or material (combining different properties), remain in the spray formed preform. Spray formed tubular products are used in chemical industrial applications, while spray formed rings of larger dimension may be used, for example, in aeroengine or powerplant turbines.
- Cylindrical billets are at present the most successful industrially realized spray formed geometries. Such billets are used, for example, as near-net shaped preforms for subsequent extrusion processes, where the advanced material properties of spray formed preforms are needed to produce high-value products. For spray forming of a billet (as shown in the figure), the spray is applied at an inclination or spray angle, mostly not directed onto the centre of the preform, but off-axis onto a rotating cylindrical substrate. The substrate is moved linearly (in most applications the substrate is moved vertically downwards, but attempts have also been made to spray billets horizontally), thereby maintaining a constant distance between the atomization and spray compaction area. To increase the mass flow rate and also to control the heat distribution in the billet properly during spray forming, multiple atomizers are used simultaneously in some applications (Bauckhage, 1997). In industrial applications, spray forming of billets in large dimensions of up to half a metre in diameter and above two metres in height, is typical.

The main advantage of spray formed materials and preforms compared to conventionally produced materials is their outstanding material properties. These can be summarized as follows:

- absence of macrosegregations;
- homogeneous, globular microstructure;
- increased yield strength;
- decreased oxygen contamination;
- good hot workability and deformability.

Literature on spray forming fundamentals and applications, and spray formed material properties, can be found in a number of specialist publications, as well as in general review journals and special conference proceedings. Material-related specific, expected or realized advantages, and properties of spray formed materials and preforms, from a research and industrial point of view, are frequently reported at the International Conferences on Spray Forming (ICSF; Wood, 1993, 1997, 1999; Leatham *et al.*, 1991). Also, the collaborative research group on spray forming at the University of Bremen edits a periodical publication, *Koll. SFB*, on research and application results of spray forming (Bauckhage and Uhlenwinkel, 1996a, 1997, 1998, 1991, 2001). In the latter, the main advantages of the different material groups produced by spray forming have been discussed:

- for copper-based alloys see Müller (1996), Hansmann and Müller (1999), as well as Jordan and Harig (1998) and Lee *et al.* (1999);
- for aluminium materials see Hummert (1996, 1999), Rückert and Stöcker (1999) and Kozarek *et al.* (1996);
- for iron-based alloys and steels see Wünnenberg (1996), Spiegelhauer and coworkers (1996, 1998) and Tsao and Grant (1999), as well as Tinscher *et al.* (1999);
- for titanium and ceramic-free super alloys see Müller *et al.* (1996) and Gerling *et al.* (1999, 2002);
- for composite materials and metal-matrix-composites (MMCs) see Lavernia (1996); and
- for lightweight materials (e.g. magnesium) see Ebert *et al.* (1997, 1998).

Presentation and exchange of scientific ideas and results of the spray forming process and the related process of thermal spraying was combined for the first time at the International Conferences on Spray Deposition and Melt atomization (SDMA), which took place in Bremen in 2000 and 2003 (Bauckhage *et al.*, 2000, 2003). Besides review papers, several papers looking specifically at microstructure and material properties, melt atomization, technical synergies from thermal spraying, process diagnostics and process analysis, new process developments, and modelling and simulation of spray forming, were presented. Select papers from these conferences have also been published in special volumes of the *International Journal of Materials Science and Engineering A* (Fritsching *et al.*, 2002).

2.2   Division of spray forming into subprocesses

From a chemical engineering and process technology viewpoint spray forming needs to be divided into a number of subprocesses. The first step is the division into the three main categories described in Chapter 1, each of which can, in turn, be further disseminated to derive requisite individual processes and process steps. A possible subdivision of the whole spray forming process is illustrated in Figure 2.3.

These subprocesses and their tasks and descriptions, as well as the aims of their modelling and numerical simulation in the frame of the integral spray forming process are described below.

(1) Melt delivery in a conventional spray forming process plant is realized either directly from the melting crucible (by a bottom pouring device, e.g. by control with a stopper rod) or more frequently by pouring the melt from the melting crucible into a tundish and from this through the melt nozzle towards the atomization area. Process instabilities caused by possible freezing of the melt in the narrow passage through the melt nozzle tip, especially in the transient starting phase of the process, are well known and feared by all industrial users of spray forming. Several attempts to improve the stability of the melt exiting from the nozzle have been made. Description and analysis of the transient melt, in terms of its temperature distribution within the tundish and its velocity profile within the nozzle help to develop suitable process operation strategies and thereby to prevent nozzle clogging.

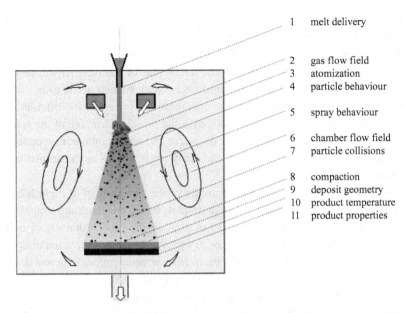

1   melt delivery

2   gas flow field
3   atomization
4   particle behaviour

5   spray behaviour

6   chamber flow field
7   particle collisions

8   compaction
9   deposit geometry
10   product temperature
11   product properties

Fig. 2.3 Subdivision of the spray forming process into subprocesses

The subsequent atomization process is mainly influenced by the temperature distribution within the melt and the dependency of its material properties upon temperature (which may change along the flow path from the crucible via the tundish until the atomization nozzle is reached). Also the metal mass flow rate may change during operation of the spray forming process due to varying temperatures and thermal conditions (or chemical reactions) in the nozzle. In addition, the local velocity distribution of the melt and the geometric dimensions of the exiting jet stream will severely influence the atomization process and the resulting melt in terms of spray properties (e.g. drop-size spectrum). Simulation of the behaviour of the melt jet stream with respect to melt/gas phase interface boundary conditions will give valuable information about this subprocess.

(2) Modelling and description of the behaviour of the gas during flow in the vicinity of the twin-fluid atomizer is needed in order to analyse melt/gas interaction properly during atomization. In spray forming, the metal melt, in most cases, is atomized by means of twin-fluid atomizers with inert gases. The main reason for using this particular type of atomizer is that these nozzles deliver high kinetic energies for fluid disintegration, in combination with high heat transfer and droplet acceleration rates in the spray by the atomization gas. In the typical range of gas pressures and gas nozzle geometries used in spray forming, gas flow in the vicinity of the atomizer is a complex dynamic process involving compressible, transonic and turbulent behaviour of the gas in the atomization area. Modelling and description of gas flow behaviour in the vicinity of the atomizer nozzle by numerical simulation is a necessary boundary condition for derivation of a melt disintegration model and is the main aim of this simulation step.

(3) A key step in the derivation of an integral spray forming process is modelling of metal melt disintegration (primary atomization). Disintegration is caused by processes that lead to instability in the fluid (melt) stream and at the interfacial phase boundary. Ongoing models take into account local interaction of the melt stream and the gas flow field and successive liquid stream disintegration (which needs to be subdivided into primary and secondary atomization, as these are governed by different physical processes). From this analysis one can derive the primary structure of the spray in its initial constellation with respect to the drop-size and drop-momentum spectrum in the spray and the overall spray geometry and spray angle as well as the droplet mass flux distribution within the initial spray.

(4) The spray structure within the spray forming process consists of in-flight accelerated, thereby cooled and partly solidified, melt droplets on the one hand and a rapidly heated (by the hot particles) and decelerated gas flow on the other. Analysis of individual droplet behaviour in the spray is a necessary precondition for derivation of droplet in-flight movement and cooling models. From here, adequate modelling and description of the droplet solidification process in the spray is achieved.

(5) Complete spray analysis and spray structure description is based on dispersed multiphase flow modelling and simulation, involving complex transport, exchange and coupling processes (for momentum and thermal energy) between the gas and the droplet phases. Such spray simulations may reveal the distribution of mass, kinetic and thermal properties of the spray at the point of impingement onto the substrate/deposit. These properties have to be derived for individual particles in the spray, as well as for the overall (integral) spray, by averaging and integration. Knowledge of spray properties is an important process variable for: prediction of the quality of the material and porosity of the sprayed product; necessary minimization of the amount of particle overspray; derivation of the shape and geometry of the sprayed product; and, finally, determination of the properties of the spray formed preform.

(6) Modelling and description of the two-phase flow inside the spray chamber allows, for example, the analysis of powder recirculation flows and possible hazardous powder pre-cipitation in the spray chamber. Fine overspray powder, which has not been compacted, swirls around the spray chamber during the spray forming process. Such powders in-fluence process control devices and measurements, and may result in the deposition of powder at hot spray-chamber walls. These recirculating powders contribute to process problems and may result in cost-intensive cleaning of the spray chamber.

Possibilities for reducing powder recirculation and overspray need to be developed. The amount by which the melt mass flow is oversprayed depends on a number of atom-ization and process conditions. Derivation, for example, of a suitable spray-chamber design based on fluid mechanical simulation and analysis, as well as adaptation of the spray-chamber geometry to the specific application and material, is a reason-able way of minimizing the overspray and preventing powder deposition at chamber walls.

(7) For high particle or droplet concentrations (dense conditions) in dispersed multiphase flow (as within the dense spray region close to the atomization area), the relevance of

particle/particle interactions, such as spray droplet collisions, increases. Binary droplet collisions in a spray may result in droplet coalescence or droplet disintegration. In both cases, the spray structure will change significantly. In spray forming applications, the different morphologies of melt droplets and particles in the spray (as there are fluid, semi-solid and solid particles in the spray) need to be taken into account, as these may exhibit very different phenomena during droplet collision. In the case of MMC spray forming, interaction of melt droplets with solid (e.g. ceramic) particles in the spray during collision needs to be analysed, e.g. to increase the incorporation efficiency of the process. These solid/liquid collision processes may result in solid particle penetration into a fluid or semi-solid melt droplet.

(8) Derivation of the fluid dynamic, thermal and metallurgical aspects of the spray forming process, as well as of those mechanisms themselves, which yield the various porosities encountered in spray formed preforms, need to be analysed in terms of droplet impact and deformation behaviour. Furthermore, analysis of fluid and semi-solid melt droplets during impact on surfaces of different morphology (these may be fluid, semi-solid or solid) is essential. Until now, the contribution of back-splashing droplets or particles that are disintegrated during impact has been poorly understood, and it is important to model their effect on the spray forming process (for example, their contribution to overspray).

(9) Based on modelling and analysis of the impacting spray and the compaction process of spray droplets, the geometry of the sprayed product needs to be derived in relation to spray process properties and deposit conditions during compaction. This will allow strategies for online process control (e.g. during temporal deviation of the process conditions from optimum) to be derived.

(10) Description of the transient material temperature and distribution of solids in the growing and cooling deposit during the spray process, and in the subsequent cooling process, may contribute a priori to analysis of the remaining product and its material properties. Thus, transient thermal analysis of the combined system, i.e. deposit and substrate, with respect to ambient flow conditions, facilitates analysis of this step of the process.

(11) Numerical models and simulations are now used in many engineering applications. Several disciplines may contribute to the development of an integral spray forming model. Classical materials science and mechanical models based on the main process conditions during spray forming and applied in the context of spray forming will contribute a priori to derivation of the remaining material properties and to identification of possible material deficits.

2.2.1   Subdivision of process steps and interaction of components

The contribution of each individual subprocess, introduced above, to the overall spray forming model, varies significantly. Several ways of combining the relevant spray forming subprocesses have already been suggested in the literature. These are introduced in the following discussion (see, for example, Lavernia and Wu (1996)).

Lawley *et al.* (1990) and Mathur *et al.* (1991) have examined the spray forming process, and have looked at how fundamental knowledge of atomization and the compaction process affect system construction.

In this sense, proper control of the main process parameters, such as substrate movement, spray oscillation, deposit temperature and so on, is essential, as shown in Figure 2.4. This diagram illustrates the relationship between independent process conditions (i.e. those con- trolled directly by the process operator), shown on the left-hand side of the figure, and the process critical conditions (i.e. those that the operator is unable to control directly). The latter include:

- the spray condition at impact, and
- the surface condition of the substrate/deposit.

The aim of Lawley *et al.*'s and Mathur *et al.*'s approaches, is to determine which parameters are process controllable. They found that, typically, the three most important parameters for characterization of the quality of the sprayed product are:

- the morphology of the deposit (deposit geometry and dimensions),
- the mass yield and the undesired metal mass loss due to overspray, and
- the microstructure of the product (porosity and grain size).

Alteration of any of these independent process variables will directly result in a change in the number of dependent process parameters and will influence the desired output properties. A more detailed diagram describing the interaction between dependent and independent process parameters, as defined by Lawley *et al.* (1990), is shown in Figure 2.5. In this diagram, five successive stages have been derived for subdivision of the spray forming process.

A number of models are required to explain the individual process steps within spray forming in order to account for the numerous cause–reaction relations exhibited at differ- ent stages, as illustrated in Figure 2.5. Lawley *et al.* (1990) have divided these into two categories:

- high-resolution models (HRM), and
- low-resolution models (LRM).

Based on the definition of these authors, high-resolution models describe the dependent variables of the process (such as temperature, velocity and solid contents of sprayed particles in the spray) that are based precisely on fundamental relations and physical conditions, and models of fluid mechanics, heat transfer and phase change during solidification with respect to the independent process variables. The low-resolution models are needed for those subprocesses and relationships where quantitative models are not available. Also, the low-resolution models may be used to average results of high-resolution models, e.g. in the form of simulation correlations. Examples of process variables and processes where high-resolution models are not available at present include: prescription of the droplet-size distribution in the spray or the problem of overspray generation. Low-resolution models are

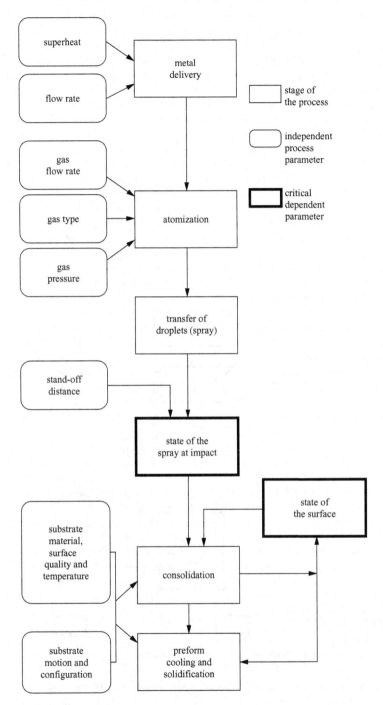

Fig. 2.4 Modelling of dependent and independent process parameters as a flow diagram (Lawley *et al.*, 1990)

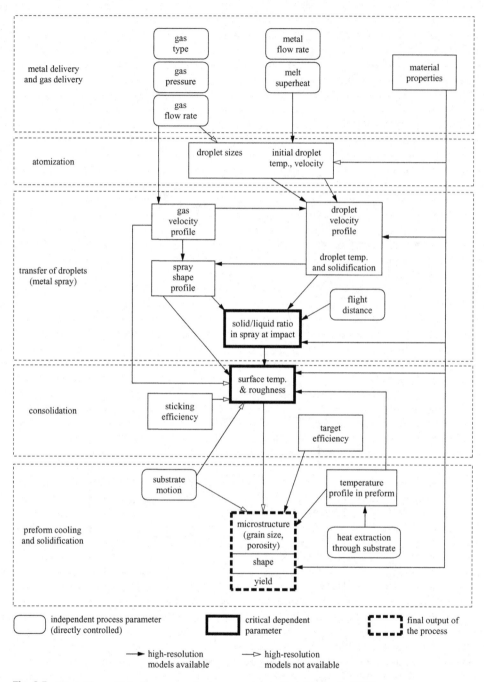

Fig. 2.5 Modelling within five successive process stages (Lawley *et al.*, 1990)

of major importance, for example, for online process control and regulation. In principle, three possible sources for derivation of low-resolution models exist:

(1) averaged results of high-resolution models, e.g. for the droplet-size distribution in the spray, or the mean temperature and the spatially averaged solidification content of the impacting spray;
(2) empirical correlations, e.g. for determining principal independent process variables (e.g. melt properties such as surface tension and melt viscosity) in relation to the mean droplet-size spectrum (of the overall spray), or the principal relation between microstructure, grain size and cooling rate of the material;
(3) empirical rules and simple logic correlations.

The present application of Lawley *et al.*'s (1990) high- and low-resolution models is also illustrated in Figure 2.5.

Through application of several low-resolution models, as defined by Lawley *et al.* (1990), the first attempt to establish an integral spray forming process model was made by Ottosen (1993). Ottosen divides the spray forming process into three main subprocesses:

(1) atomization and disintegration of the fluid (melt);
(2) interaction of gas and droplet/particles in the spray; and
(3) droplet impingement, compaction and interaction of the particles with an underlying surface.

In his work, Ottosen (1993) defines modelling and simulation of spray development, and the complex momentum and heat transfer and exchange processes within the spray, as key to the analysis of an integral spray forming model. He also describes structure formation during the deposit growth and modelling of the temperature distribution in the deposit and the substrate, located below the deposit, as an important subprocess that dominates the overall performance of the spray forming process. Based on Ottosen's modelling approach, Pedersen (2003) derived a coupled integral process model for spray forming based on:

• an atomization model,
• a shape model, and
• a deposition model.

The main emphases of this coupled approach are analysis of the surface temperature of the spray preform and derivation of an empirical correlation to be used, for example, in process control (see Chapter 7).

Modelling of physical subprocesses is also a necessary condition for optimum *in-situ* and online control and regulation of the spray forming process.

Bauckhage and Uhlenwinkel (1996b) describe the possibilities of an automated and optimized spray forming process, based on the division of spray forming into three subprocesses:

(1) melting and atomization,
(2) particle transport in the spray, and
(3) compaction.

All visually observable, measurable and controllable material properties, the preform geometry and the process conditions are then related to these three subprocesses. In their analysis, the qualitative influence of changing process conditions during spray forming within each of these individual subprocesses, and the effect that this has on the overall process and resulting properties, are discussed. The analysis shows that control of the particle size spectrum in the spray is the most important parameter during spray forming, and that this, in turn, can be readily influenced by regulation of the mass flow rate of the atomizer gas (via the atomization gas pressure, at constant thermal conditions). The aim of process control is to be able to influence atomization and, immediately connected with atomization, to control the multiphase transport processes in the spray, which build the adjustable parameters of the spray forming process. Fluctuation of the process conditions, e.g. melt properties may alter due to temperature variations or melt delivery fluctuations, may disturb the process parameters. Also, melt mass flow rate variations from the tundish during the process, related to chemical reactions or thermal variations, need to be controlled. Continuous measurement of the melt mass flow rate and online comparison of actual and rated values during the process are therefore essential. In order to compare actual and rated melt mass flow values steadily, when the measured value deviates from the rated value, one has to regulate the process by changing the gas pressure (as it is impossible to alter the melt flow rate directly). Therefore, in order to maintain a constant gas-to-melt ratio (GMR), the gas pressure, and thus the gas flow ratio, needs to be changed. The GMR directly influences the particle size distribution in the spray. Thus to keep it constant during the process (or to change it in a prescribed way) the gas pressure needs to be changed, especially when the melt flow rate is altered by other process instabilities.

An empirical spray forming process model, describing the relationship between process parameters and the resulting product quality, has been introduced by Payne *et al.* (1993). As a suitable base for process control, modern sensors and neural networks have been introduced for monitoring purposes. Based on a small number of test runs, a neural network has been trained for analysis and control of the spray forming process for production of tubular elements. The main measurement parameters are: the exhaust gas temperature; estimation of the surface roughness; and porosity of the sprayed product during the process. Control parameters to be measured and controlled are: the GMR, which is dependent on the atomizer gas pressure; the melt flow diameter; and the external spray-chamber pressure. In addition, the withdrawal rate and the rotational speed of the tube, the spray height and the melt temperature are also measured. For suitable process control, Payne *et al.* (1993) have identified:

- directly controllable process parameters: e.g. spray time, melt temperature and GMR;
- indirectly controllable process parameters: e.g. exhaust gas temperature, deposit surface roughness and porosity.

Rebis *et al.* (1997) introduced a simulation program aimed at the development of a production rule and control strategy for manufacturing of spray formed products whose geometries are based on computer-aided design (CAD) and upon the material properties of the alloy used. Therefore, four modelling segments are introduced:

(1) a planning system based on neural networks that connects the input parameters to resulting product quality;
(2) a generation program based on step (1) that derives, based on the plant parameters (including robot data and spray parameters), the operational parameters of the process;
(3) a module for process simulation based on the work on process modelling at Drexel University (Cai, 1995), which aims to prescribe the three-dimensional geometry of the spray formed product (based on spatial and temporal averaged-distribution of the mass flux of the particles in the spray, the compaction efficiency and spatial movement of the spray and substrate);
(4) a graphical tool for model and process visualization.

Evaluation of a relatively simple process modelling experiment for development of a more complex process simulation model, describing the spray conditions within the spray forming process in relation to the remaining porosity inside the sprayed deposit, has been described by Kozarek et al. (1998). These experimental models, their derivation and describing process parameters, have been chosen based on simulations of the mean solidification ratio of impinging spray droplets. Process parameters which have been taken into account are:

- the atomizer gas pressure;
- the GMR;
- the spray distance between atomizer and substrate;
- the mean droplet velocity, and the mean droplet size; as well as
- melt superheating.

The operation window for minimization of the porosity has been derived from the model, based on a transient particle tracking within the spray. Based on the results of the model, proposals for scaling-up spray forming process into larger dimensions have been made. These authors identify the thickness of the mushy layer (semi-solid area) at the top of the deposit as a key process control mechanism. A model-based algorithm is proposed for this control process.

Bergmann et al. (1999a,b, 2000) use numerical modelling and simulation to determine the operational mechanisms of individual process and operational parameters affecting thermal behaviour within pure copper (Cu) and steel (C30 and C105) spray forming processes. They divide the spray forming process of metallic preforms into:

- melt flow in the tundish,
- spray flow, and
- compaction and cooling of the deposit.

Bergmann and coworkers combine these subprocesses to defined an integral model. By varying model parameters, common properties and their relationships during the thermal behaviour of the material are obtained. The starting point for the investigations is the hypothesis that, especially during spray forming of high-volume products such as billets, residence time variations of material elements at different temperature levels within the spray formed product (and the related distribution of the solidification velocity of different material elements) result in an inhomogeneous distribution of properties in the material. For example, if the cooling velocity is too low, recrystallization processes may occur and thereby grain coarsening will result, which is an undesired quality of spray formed products (Jordan and Harig, 1998). In addition, spatial distribution of the cooling behaviour of billets may cause thermal and residual stresses that may result in hot cracks inside the deposit. Commercial fabricators of spray formed billets are aware of such problems. A straightforward control mechanism to improve the spray forming process has not been identified yet. A common phenomenon during spray forming of copper is the establishment of high-volume areas within the product of increased porosity, and are called 'cauliflowers' from visual observation of their macrostructure (Jordan and Harig, 1998). For description and analysis of the transient behaviour of the thermal history of spray formed products, Bergmann (2000) derived (based on the work of Zhang (1994)) an integral thermal model of the spray forming process. This model describes the thermal history of a melt element. This analysis of thermal history begins with the temperature level in the tundish, via the flow in the tundish, to the atomizer nozzle; and, next, look at the solidification behaviour of melt droplets in the spray down, at the impact onto the deposit and at subsequent consolidation and cooling within the sprayed product. Results from each preceding process step are transferred to the model of the next process step as a boundary condition (see Chapter 7).

Chapter 2 illustrates that it is possible to derive a general model for spray forming, which most researchers have subdivided into three main subprocesses. A more detailed division into eleven steps, as examined in Section 2.2, is possible and provides a somewhat more general and individual assessment from a process technology viewpoint. This subdivision is based on the specific research areas where most of the work has been conducted. A somewhat different subdivision is possible and, in other spray applications, sometimes even strongly desired (based on the aim of the model and the expected precision of the results). Despite the different viewpoints, recognition of the common characteristics of all spray forming processes is the main reason for modelling the multiphase flow properties correctly. This is the main focus of this book.

3 Modelling within chemical and process technologies

The successive steps involved in spray modelling and simulation within chemical and process technologies may be characterized by a single common scheme, which is illustrated in Figure 3.1.

Physical model

Having divided the process or system into modular subprocesses (see Section 2.2), the starting point for any modelling procedure is the identification and derivation of the main physical mechanisms of each individual process step. For this derivation, balances of physical properties are to be performed which, for example, yield conservation equations for the main physical transport and exchange parameters (normally in terms of ordinary and partial differential equations) involved. For analysis of the main multiphase flow properties involving heat transfer and phase change within atomization and spray processes, the general continuum mechanics conservation and transport equation formulation can be written as:

$$
\frac{\partial}{\partial t}(\rho \Phi) + \frac{\partial}{\partial x}(\rho u_g \Phi) + \frac{1}{r^\gamma}\frac{\partial}{\partial r}(\rho r^\gamma v_g \Phi) - \frac{\partial}{\partial x}\left(\Gamma \frac{\partial \Phi}{\partial x}\right) - \frac{1}{r^\gamma}\frac{\partial}{\partial r}\left(r^\gamma \Gamma \frac{\partial \Phi}{\partial r}\right)
$$
$$
= S_\Phi - S_{\Phi_p}, \tag{3.1}
$$

where Φ is the general transport variable, Γ is the diffusivity of the transport value and S is a synonym for the related source or sink terms . This conservation equation reflects the influence of transient behaviour, convective and diffusive transport, as well as production and dissipation of the transported variable Φ. The source terms in the present context can be referred to as follows. The source S_Φ identifies the inner phase sources or sinks (e.g. pressure) and the term S_{Φ_p} identifies those sources that result from coupling of the continuous phase to the dispersed phase in two-phase flow. This general transport equation is valid within a two-dimensional plane $(x-y)$ or cylindrical $(x-r)$ coordinate system. For cylindrical coordinates, $\gamma = 1$; and for Cartesian plane coordinates, $\gamma = 0$.

In addition, for common flow simulations an appropriate turbulence model is used. In this book, mainly two-dimensional processes are studied. Extension of the proposed methods into three dimensions is possible, but because of the great computational power that is needed, extension in most cases is not desirable. Therefore, for analysis of two-dimensional, single or multiphase turbulent continuum flows with heat transfer, the well-established k–ε model for turbulence behaviour (Launder and Spalding, 1974) is mostly used.

Table 3.1 *Terms used in the general conservation equation (3.2)*

Φ	S_Φ	$S_{\Phi P}$	Γ
1	—	—	—
u_g	$\dfrac{\partial}{\partial x}\left(\Gamma \dfrac{\partial u_g}{\partial x}\right) + \dfrac{1}{r^\gamma}\dfrac{\partial}{\partial r}\left(\Gamma r^\gamma \dfrac{\partial v_g}{\partial x}\right) - \dfrac{\partial p}{\partial x} + \rho g_x$	$S_{p,x}$	μ_{eff}
v_g	$\dfrac{\partial}{\partial x}\left(\Gamma \dfrac{\partial u_g}{\partial r}\right) + \dfrac{1}{r^\gamma}\dfrac{\partial}{\partial r}\left(\Gamma r^\gamma \dfrac{\partial v_g}{\partial r}\right) - \dfrac{\partial p}{\partial r} - 2\Gamma^\gamma \dfrac{v_g}{r^2}$	$S_{p,r}$	μ_{eff}
h_g	—	S_h	$\dfrac{\mu_{\text{eff}}}{\sigma_h}$
k	$G_k - \rho\varepsilon$	—	$\dfrac{\mu_{\text{eff}}}{\sigma_k}$
ε	$\dfrac{\varepsilon}{k}(c_1 G_k - c_2 \rho\varepsilon)$	—	$\dfrac{\mu_{\text{eff}}}{\sigma_c}$

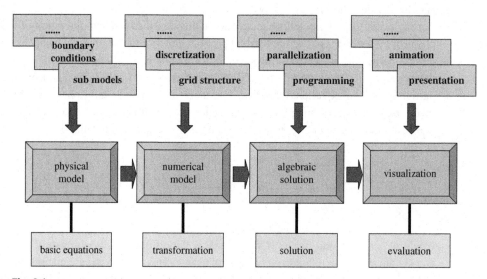

Fig. 3.1 Flow diagram of modelling of chemical engineering processes

Therefore, in summary, the terms in the general conservation equation (3.2), are defined in Table 3.1.

In this context, the transport variables are: $\Phi = 1$ for mass conservation, u_g and v_g are the velocity components for momentum conservation, h_g is the conservation of specific enthalpy of the liquid for thermal energy conservation, k is the turbulent kinetic energy and ε is the dissipation rate of the turbulent kinetic energy. The source term G_k in the turbulence conservation equation of k is formulated as:

$$G_k = \mu_{\text{eff}}\left\{2\left[\left(\frac{\partial u_g}{\partial x}\right)^2 + \left(\frac{\partial v_g}{\partial r}\right)^2 + \left(\frac{\gamma v_g}{r}\right)^2\right] + \left(\frac{\partial u_g}{\partial r} + \frac{\partial v_g}{\partial x}\right)^2\right\}. \tag{3.2}$$

The effective viscosity μ_{eff} (total turbulent viscosity based on a Boussinesq approximation) is described in the k–ε model by the sum of the (material dependent) viscosity μ and the (flow-field-dependent) turbulent viscosity μ_t:

$$\mu_{\text{eff}} = \mu + \mu_t = c_\mu \rho \frac{k^2}{\varepsilon}. \tag{3.3}$$

The standard coefficients of the k–ε model are:

c_μ	c_1	c_2	σ_k	σ_ε	σ_h
0.09	1.44	1.92	1.0	1.3	0.9,

where the first five coefficients are within the turbulent kinetic energy and dissipation rate equations and σ_h is the turbulent Prandtl number, relating the turbulent diffusive transport of thermal energy to viscous diffusive transport.

The classical, well-known conservation equations found in textbooks (see e.g. Bird, Stewart and Lightfoot, 1960) sometimes need to be completed by specific submodels for description of specific flow conditions or physical subprocesses. In the present context, for example, the source terms for phase change during solidification of the metal melt need to be adequately formulated and added. For proper transfer of the modelling approach to other processes, these submodels need to be derived and formulated in the most general way. In most cases, such submodels are derived from theoretical approaches in connection with the results of well-defined experimental investigations; and, more recently have been derived from detailed (direct) numerical simulations.

Another very important part of a physical process model is the formulation of suitable initial (starting) and boundary conditions for the specific conservation variables, which need to be properly fitted to the process under investigation.

Numerical model

Within simulations, transfer of the fundamental conservation equations describing the process into a system of algebraic equations that can be solved by mathematical methods is the task of numerical mathematics. Efficient algorithms and solution strategies need to be applied, which keep in mind the subsequent computer-based algebraic solution of the set of equations. Most popular and well-distributed discretization processes in the frame of single- or multiphase continuum flows are based on discretization of the differential equations in finite differences, finite volumes or finite elements. Several possibilities exist to discretize conservation equations properly in order to conserve their contents and guarantee numerical accuracy and stability. The different methods of discretization and implementation of numerical models will not be introduced in detail here. Introduction and comparison of numerical models in the context of the physical modelling to be done here is documented in fundamental literature:

- for example, with respect to the fundamentals of fluid dynamics simulation (or computational fluid dynamics, CFD) in the contributions of Fletcher (1991) or Ferziger and Peric (1996);

- for application of numerical methods for computing fluid flow in boundary fitted coordinate systems see Schönung (1990);
- for simulation of heat and momentum transport processes see the fundamental work of Patankar (1981);
- in the area of dispersed multiphase flow see Crowe *et al.* (1998), as well as Sommerfeld (1996);
- for some applications of simulation within atomization modelling of metal melts see Yule and Dunkley (1994) and Dunkley (1998).

In the area of numerical simulation and the solution of fundamental conservation equations, the role of commercial, or freeware and shareware programs (to be distributed, for example, via the internet), has tremendously increased in recent years. Relevant information on fundamentals, literature, simulation programs and other information on simulation of fluid flow (CFD) can be found, for example, at the excellent and informative homepage of Chalmers University at: www.cfd.chalmers.se/CFD-Online/ (see Useful web pages, p. 269).

The main advantages of using so-called multipurpose or specifically designed commercial programs are the rapid access and (mostly) easy usage of these programs. Hereby, several technical processes may be analysed with quick success. The development of such program systems is based on a great amount of knowledge and experience as well as manpower. Such commercial programs are suitable for a number of applications and boundary conditions where the amount of calculations necessary is discouraging. However, the disadvantage of most commercial simulation programs is their 'black-box character'. Often, the modelling fundamentals, solution strategies and algorithms are not well described and documented. Missing, or difficult to derive, extensions of such programs to one's own submodels, make application for specific boundary conditions and purposes hard to achieve and limits the application of such codes. The term general-purpose code also often means that commercial programs are not up-to-date within specific applications, processes and developments. Sometimes it is even recommended that program developers do not adopt each proposed trend, as new model developments need to be tested carefully, based on a number of reference cases, before being introduced into a general code. However, programs for specific processes are mostly developed in universities and research institutes, and therefore have better application. The pros and cons of application of commercial software for simulation purposes is fundamental, but needs to be evaluated from application to application. Some guidance by experienced modellers, simulation program users and developers is needed in the decision process to find a suitable simulation program. It needs to be said that commercial program systems in no way lead to better or worse results than specific in-house codes developed on one's own. Both approaches are, generally, based on identical conservation laws and numerical methods. The accuracy of results is mostly influenced by taking account of all necessary physical preconditions and boundary conditions of suitable submodels. Often, relevant physical subtasks are simplified or neglected or not taken into account, simply because of lack of data, or no models are available, or the specific models do not sufficiently represent the physical behaviour to be described. The numerical and computational efficiency of commercial programs is, in most cases, better than that of specifically developed in-house codes.

In this book, several kinds of computer codes and programs will be described and have been used for simulation. Commercial programs for analysis of multiphase flow behaviour are used, such as PHOENICS (Rosten and Spalding, 1987), FLOW3D (now called CFX; Computational Fluid Dynamics Services, 1995), RAMPANT (Rampant, 1996), FLUENT (creare.x Inc, 1990), STAR-CD (Computational Dynamics Ltd, 1999), FLOW-3D (Flow Science, 1998) and others in different versions. Also, specifically developed and programmed in-house codes from the author's research group for simulation of dispersed multiphase flow and heat transfer problems have been used (see Useful web pages, p. 269). The transition between both types of codes and simulation approaches is floating.

Algebraic solution

The result of the numerical model is a set of (in most cases linear) algebraic equations. This huge set of equations needs to be approached and solved by efficient numerical equation solvers (see e.g. Barrett *et al.*, 1994). The resources necessary to handle such equation systems are increasingly given by computer power, but can also be determined by distributing the algebraic solution onto several processors within a parallel or cluster machine. In order to apply parallel solution strategies, suitable numerical discretization schemes need to be applied in the foregoing numerical modelling step. In addition to classical parallel machines, concepts of workstation or PC clustering come into focus nowadays, e.g. based on the Beowulf concept (Sterling *et al.*, 1999).

Visualization

The enormous amount of data derived as a result of the algebraic solution represents the distribution of all relevant variables at discrete grid points within the numerical grid system. These huge data sets need to be filtered and properly condensed in order to derive user-dependent and relevant properties of the simulated physical process. This simulation step is called post-processing. Here, two- or three-dimensional methods and transient animation and data visualization may be used. Without sophisticated visualization tools, adequate evaluation of the huge data sets cannot be realized (see Baum, 1996) for engineering purposes. Commercial tools are available in this area also, which have been especially developed for visualization of numerical simulation data (see Useful web pages, p. 269).

4 Fluid disintegration

Having divided the atomization and spray process modelling procedure into three main areas:

(1) atomization (disintegration),
(2) spray, and
(3) compaction,

in this chapter we look specifically at the disintegration process as it applies to the case of molten metal atomization for spray forming. We begin by breaking down disintegration into a number of steps:

- the melt flow field inside the tundish and the tundish melt nozzle,
- the melt flow field in the emerging and excited fluid jet,
- the gas flow field in the vicinity of the twin-atomizer,
- interaction of gas and melt flow fields, and
- resulting primary and secondary disintegration processes of the liquid melt.

Several principal atomization mechanisms and devices exist for disintegration of molten metals. An overview of molten metal atomization techniques and devices is given, for example, in Lawley (1992), Bauckhage (1992), Yule and Dunkley (1994) and Nasr *et al.* (2002). In the area of metal powder production by atomization of molten metals, or in the area of spray forming of metals, especially, twin-fluid atomization by means of inert gases is used. The main reasons for using this specific atomization technique are:

- the possibility of high throughputs and disintegration of high mass flow rates;
- a greater amount of heat transfer between gas and particles allows rapid, partial cooling of particles;
- direct delivery of kinetic energy to accelerate the particles towards the substrate/deposit for compaction;
- minimization of oxidation risks to the atomized materials within the spray process by use of inert gases.

A common characteristic of the various types of twin-fluid atomizers used for molten metal atomization is the gravitational, vertical exit of the melt jet from the tundish via the (often cylindrical) melt nozzle. Also, in most cases, the central melt jet stream is surrounded by gas flow from a single (slit) jet configuration or a set of discrete gas jets, which flow in a direction parallel to the melt flow or at an inclined angle to the melt stream. The coaxial

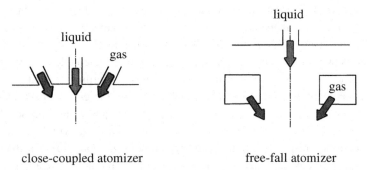

close-coupled atomizer   free-fall atomizer

Fig. 4.1   Free-fall and close-coupled atomizer for metal powder generation and spray forming

atomizer gas usually exits the atomizer at high pressures with high kinetic energy. Two main configurations and types of twin-fluid atomizers need to be distinguished for molten metal atomization: i.e. the confined or close-coupled atomizer, and the free-fall atomizer. Both are illustrated in Figure 4.1. The gas flow in the close-coupled atomizer immediately covers the exiting melt jet. Within the confined atomizer the distance between the gas exit and the melt stream is much smaller than in the free-fall arrangement, where the melt jet moves a certain distance in the direction of gravity before the gas flow impinges onto the central melt jet. The close-coupled configuration generally tends to yield higher atomization efficiencies (in terms of smaller particles at identical energy consumption) due to the smaller distance between the gas and melt exits. However, the confined atomizer is more susceptible to freezing of the melt at the nozzle tip. This effect is due to extensive cooling of the melt by the expanding gas, which exits in the close-coupled atomizer near to the melt stream. During isentropic gas expansion the atomization gas temperature is lowered (sometimes well below 0 °C). Close spatial coupling between the gas and melt flow fields contributes to rapid cooling of the melt at the tip of the nozzle. This freezing problem is especially relevant for spray forming applications, as discontinuous batch operation is a standard feature of all processes (e.g. as a result of batchwise melt preparation or the limited preform extend to be spray formed). The operational times of spray forming processes range from several minutes up to approximately one hour. The thermal-related freezing problem is most important in the initial phase of the process, when the melt stream exits the nozzle for the first time. At this point in the operation, the nozzle tip is still cool and needs to be heated, for example, by the hot melt flow. Due to the time required to heat the nozzle, thermal-related freezing problems are often observed in the first few seconds of a melt atomization process.

In addition to the problem of thermal-related freezing within the nozzle, chemical or metallurgical problems in melt delivery systems are frequently found. Many of these problems are still to be solved in melt atomization applications. A range of problems arises from a possible change in composition of the melt, or that of the tundish or nozzle material, due to possible melt/tundish reactions or melt segregational diffusion effects. The reaction kinetics arising from this type of behaviour is somewhat slower than in the thermal freezing process, and may contribute to operational problems at a later stage.

Free-fall atomizers are much less problematic than close-coupled atomizers in terms of thermal freezing processes, as the melt jet stream and gas stream are well separated at the exit of the melt from the delivery system (tundish exit). Therefore, cooling of the melt due to cold gas occurs at a later point in the system than within close-coupled atomizers. An additional advantage of the use of a free-fall atomizer in spray forming, is that controlled mechanical or pneumatic scanning is possible and, therefore, the gas atomizer oscillates with respect to one axis. Also, the spatial distribution of gas jets within the nozzle can be modified during atomization. By doing so, the free-fall atomizer provides an additional degree of freedom with respect to control and regulation of the mass flux distribution of droplets in the spray. In other atomizer nozzle systems, this important physical property of the spray can (within a running process) only be influenced by changing the atomizer gas pressure. By controlled scanning of the nozzle, the mass flux can also be distributed over a certain area (necessary, for example, for flat product spray forming).

4.1   Melt flow in tundish and nozzle

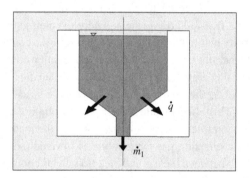

In spray forming, the melt is delivered continuously to the tundish and the melt nozzle. The exit velocity of the metal melt depends on the pressure in the tundish (mostly ambient) and therefore on the metallostatic height of the melt in the tundish only (for a predefined material density). Therefore, the melt mass flow rate depends on the size of the nozzle exit, the density of the melt and the exit velocity of the melt. The metallostatic height is governed and kept constant within the process by automatic control procedures or manual observation and regulation. Analysis of the liquid flow field inside an atomization nozzle is an important condition for derivation and assessment of the disintegration model during atomization (see Löffler-Mang, 1992). Within spray forming, investigation of the exit flow condition of the melt also derives information about:

- conditions that may lead to thermal-related freezing of the melt within the nozzle,
- the geometry of the melt stream and its kinetic and thermal state during flow from the melt nozzle tip downstream into the atomization zone below the atomizer.

4.1.1   Melt flow inside the nozzle

Besides nozzle geometry, the most important input parameters required to model tundish flow are the mass flow rate and the temperature of the melt. The mass flow rate, related to the metallostatic height of the melt within the tundish, is given as:

$$\dot{m}_l = \varphi A_{\text{out}} \rho_l \sqrt{2gh}, \tag{4.1}$$

where φ is the friction or velocity number of the nozzle, describing losses due to friction and vortex formation. In experimental investigations of the flow behaviour in spray forming applications, for typical tundish and nozzle systems with model fluids, Bergmann (2000) found that the velocity number depends on the Reynolds number only:

$$\varphi = 1 - \frac{7.96}{\sqrt{\text{Re}}}. \tag{4.2}$$

From this correlation, it can be deduced that, during spray forming and atomization, the mass flow rate of a steel melt passing through a typical tundish, e.g. at a typical Reynolds number of Re = 10 000, can deviate by as much as −10% from the ideal (frictionless) value. Deviations from inviscid theory of that order are typically found during experimental investigations on spray forming processes (Bergmann, 2000).

Besides mass flow rate and mean velocity, the local flow structure of the melt in the tundish and at the melt exit from the tundish, as well as during free-fall flow into the atomization area, is of importance, as it controls some of the main features of the atomization process and its results (e.g. the drop-size distribution in the spray). During flow of the melt into the atomization area, an important parameter is the changing geometrical dimension of the melt jet or sheet (the latter, for example, also applies to swirl pressure atomizers, see Lampe, 1994). In addition, the velocity distribution and turbulence structure within the melt jet stream contribute to wavelength interactions between the melt and gas that initiate the disintegration process.

In spray forming, knowledge of the temperature and velocity distribution at the exit nozzle of the melt is of special interest, as discussed previously. This is the region where thermal freezing may occur. In order to prevent freezing, superheating in the crucible, prior to pouring of the melt into the tundish, can be considered. The temperature to which the melt must be heated is evaluated from the heat loss of the melt inside the tundish and at the melt nozzle. To calculate the fluid and heat flow dynamics of the melt, simultaneously with the heat conduction process within the solid material of the tundish and the nozzle wall, the so-called conjugate heat transfer problem needs to be solved. In this way, the transition from heat conduction in the solid material to heat transport in the adjacent liquid is not coupled to a predetermined heat transfer coefficient. The derivation of the heat transfer value across the wall is part of the numerical solution of the coupled problem. As boundary conditions, only the continuous temperature distribution (no jump) and matching of the heat fluxes on both sides, within the inner side (fluid) and the outer side (wall), need to be prescribed. In the case of laminar flow behaviour, this can be done by:

$$\dot{q}_i = \lambda_i \left. \frac{dT}{dn} \right|_i = \lambda_a \left. \frac{dT}{dn} \right|_a = \dot{q}_a. \tag{4.3}$$

For calculation of the conjugate heat transfer problem in turbulent flows, appropriate boundary conditions (wall functions) dependent on the turbulence model used need to be applied to the fluid side. These are normally prescribed in terms of logarithmic wall functions.

Exact fulfilment of all physical boundary conditions cannot often be realized during modelling and simulation. For successful calculation, simplifying approximations and boundary conditions need to be derived and applied. The influence of inexact representation of boundary conditions on the result of the calculation must be negligible or small, or should be unimportant for the aim of the investigation. At the very least, the influence of simplifications has to be carefully evaluated, as exactly as possible. In the present case, such simplifying assumptions and boundary conditions for the tundish simulation are:

- Only steady-state cases of fluid flow and heat transport and transfer will be regarded.
- Material properties are assumed to be constant within the range of varying temperatures. A steel is taken as the standard melt material, and its temperature at the free surface (upper boundary of the tundish model) is assumed to be constant at a superheating of 50 °C above the specific melting temperature. At the upper geometric boundary (tundish top), a constant temperature profile of the inflowing melt, not varying with radius, is assumed.
- Possible solidification effects of the melt by cooling below the solidus temperature (e.g. on melt viscosity) are not taken into account; only the amount of subcooling will be analysed; and the melt is always regarded as fluid.
- Flow is assumed to be in a laminar state. Highest local Reynolds numbers of the melt flow are achieved at the tip of the tundish exit (at the point of the smallest flow area). Here maximum Reynolds numbers of up to $Re = 10\,000$ may be achieved, but only at a very small length scale, where laminar/turbulent transition of the flow is not realistic.

  In an investigation by van de Sande and Smith (1973) on the influence of nozzle geometry on the critical intake length of fluid jets during pressure atomization, the critical Reynolds number for transition from laminar to turbulent flow is:

$$Re_{crit} = 12\,000 \left(\frac{l}{d} \right)^{-0.3}. \tag{4.4}$$

  This critical Reynolds number is not reached in all flow situations and materials within spray forming.

- In the example of tundish modelling presented here, simplified tundish geometry is used, derived from approximated tundish geometries in the Bremen trial spray forming plants. The main geometrical parameters of this model are illustrated in Figure 4.2 and listed in Table 4.1.
- The influence of the cooling effect due to primary gas flow on the melt stream in the conical part of the tundish exit is taken into account in an integral approach. Therefore, at

Table 4.1 *Geometry of the tundish*

d_1 [mm]	d_2 [mm]	h_1 [mm]	h_2 [mm]
4	28	5	90

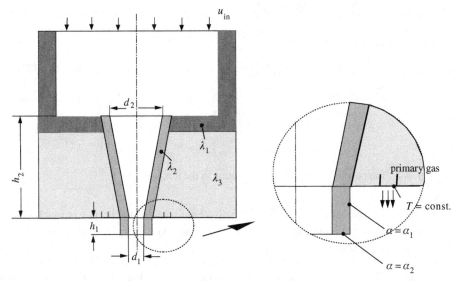

Fig. 4.2 Boundary conditions for modelling the flow in the tundish

the lower radial circumference of the tundish (upper cylindrical area), adiabatic boundary conditions are assumed.

- The gas flow is rapidly cooled due to its expansion from the high prepressure in the plenum to the exit pressure within the primary gas flow nozzles. Based on the assumptions that the gas has ambient temperature in the primary gas plenum prior to expansion and that the expansion process is described by an isentropic change of state of the gas:

$$T_g = T_u \left(\frac{p_0}{p_u} \right)^{\frac{\kappa-1}{\kappa}} ,$$

(4.5)

the temperature of the gas may achieve levels well below 0 °C. It is assumed that the specific section of the wall directly at the primary gas exit will have the same temperature as the gas flow in that region.

- At the free tip of the tundish, constant heat transfer coefficients to the gas are assumed as $\alpha_1 = 500$ W/m² K and $\alpha_2 = 100$ W/m² K.
- The materials comprising the conical exit tube of the tundish and the tundish itself are identical, and are made from the same ceramics (e.g. Al_2O_3), having a heat conduction coefficient of $\lambda_1 = \lambda_2 = 2.3$ W/m K. The tundish is embedded into an isolation layer made of graphite–hard felt with a heat conduction coefficient of $\lambda_3 = 0.25$ W/m K.

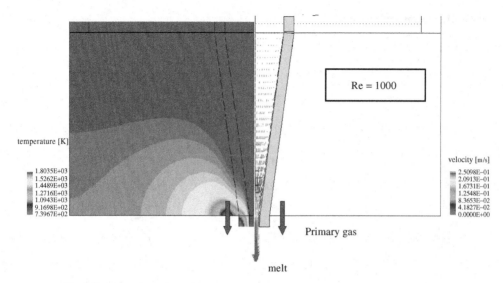

Fig. 4.3 Temperature and velocity distribution in the tundish

- At the free tip of the tundish, convective heat transfer to the gas is much greater than heat loss due to radiation (for an assessment see Andersen (1991)). Therefore, radiation is neglected at the free boundary.

An example of the results calculated from the above tundish model is shown in Figure 4.3. Here only a magnified view of the conical part of the tundish tip is to be seen. On the left-hand side of the figure the temperature distribution is shown as isolines inside the fluid and the solid material. The right-hand side of Figure 4.3 illustrates the calculated velocity distribution of the melt in vector format. This calculation has been done for Re = 1000 (with respect to the state at the tundish tip Re = $u_a(d_1/\nu_1)$, where the mean exit velocity u_a is averaged from the flow rate). It can be seen that the primary gas flow acts as the main thermal energy sink in that system and that great temperature gradients may result at the tip of the tundish.

Normalized velocity profiles of the melt flow at the tundish exit for different Reynolds numbers are illustrated in Figure 4.4. Normalization of the velocity profiles is done through the volumetric-averaged velocity for each Reynolds number. The influence of wall friction and the establishment of boundary-layer-type flow in the vicinity of the wall are small in the spray forming range of interest for Re > 5000. Here the velocity profile is almost flat in the main region in the centre of the tundish exit tube. The theoretical length over which a fully developed parabolic velocity profile is achieved during tubular flow is:

$$l/d = A\, \mathrm{Re}^B, \tag{4.6}$$

where the coefficients are $A = 9.06$ and $B = 1$ (Truckenbrodt, 1989). From the length/diameter aspect ratio of the melt tube exit in the present case ($l/d = 1.25$), it can be stated that fully developed velocity profiles may be achieved for small Reynolds numbers only (Re < 20).

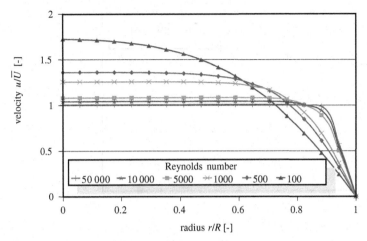

Fig. 4.4   Velocity distribution in the tundish tip

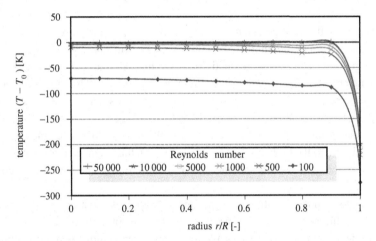

Fig. 4.5   Temperature distribution in the tundish tip

Figure 4.5 shows the calculated temperature distribution of the melt at the tundish exit together with the calculated wall temperatures for different Reynolds numbers. Obviously, the thermal energy content of the melt is so high that remarkable cooling effects can be determined only for very small Reynolds numbers.

With respect to the spray forming process it needs to be mentioned that the discontinuous batch process in which spray forming is operated, especially in the transient phases (which frequently occur in the beginning of each spray run before a steady-state temperature level is achieved and in the end phase of each spray run when the pouring crucible is emptied), leads to small exit velocities of the melt and therefore small Reynolds numbers. In spray forming practice, the starting phase of each spray forming run is typically initiated without any atomizer gas flow in order to achieve sufficient heating of the tundish walls from the poured melt. Before the atomization process begins, by adding gas to the process and thereby

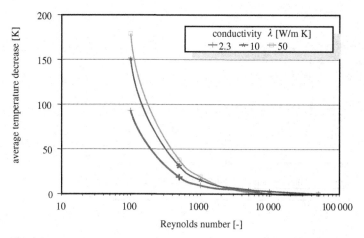

Fig. 4.6 Mean temperature in the tundish for different melt materials

causing an intense cooling in the vicinity of the nozzle, a sufficiently high Reynolds number is reached.

The dependence of the averaged melt cooling rate, between the inflow into the tundish and the exit from the bottom of the tundish (averaged over the flow area), on Reynolds number is shown in Figure 4.6. Here, heat conductivity of the tundish material λ_1 is chosen as the variable parameter under investigation. Using tundish materials with higher conductivity, the assumed superheating of the melt is sufficient for Reynolds numbers Re > 200, which result in melt temperatures at the tundish exit which still exceed the solidus temperature of the melt. Because of flat temperature profiles and small radial temperature gradients within the flowing melt (due to the high conductivity of the melt), the averaged temperature across a cross-section of the melt is a suitable measure for estimation of the freezing potential of the melt.

Comparison of simulated results with experimental values is often necessary for validation and to check the accuracy of the modelling approach. In addition, comparison with classical analytical solutions may be helpful for the interpretation of model simulation results. In the aforementioned problem of melt flow cooling in the tundish, the modelling results in the tundish tip can be compared to fully developed thermal flow in an ideal hydrodynamic tube, within a cylindrical segment 5 mm in length (the end of the tundish). It is assumed that cooling occurs over the outer tube walls. Thermal balancing results in a differential equation for the temperature distribution (Anderson, 1991):

$$-\frac{\dot{m}c_p}{2\pi r_a}\frac{dT}{dx} = \alpha_a\left[T + \dot{m}c_p\frac{\ln(r_a/r_i)}{2\pi\lambda}\frac{dT}{dx} - T_u\right]. \tag{4.7}$$

Analytical solution of Eq. (4.7) gives:

$$T(x) = (T_{in} - T_u)\exp\left[\frac{-2\pi x}{\dot{m}c_p\left(\frac{1}{\alpha_a r_a} + \frac{\ln(r_a/r_i)}{\lambda}\right)}\right] + T_u. \tag{4.8}$$

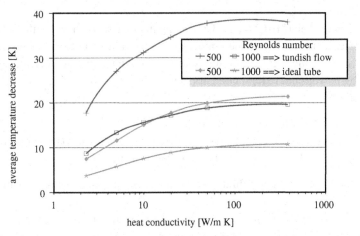

Fig. 4.7 Mean temperature in the tundish: comparison to theoretical solution

Comparison of analytical and numerical solutions is done in Figure 4.7 for two different Reynolds numbers. It can be seen that the simulated averaged cooling rate of the melt for the whole tundish segment is twice as high as the cooling rate calculated from the analytical solution, which is valid within the short cylindrical tip element of the pouring tube only, while the simulation covers the whole tundish area.

4.1.2   Fluid jet

Analysis of the two-fluid flow field before atomization of the melt (coaxial flow of gas and melt) illustrates a flow situation with a relatively small phase boundary (interface) between gas and melt (in comparison to the large phase boundary in the dispersed two-phase flow within the spray after atomization). This area is of special interest for those atomizers that transport and deform the fluid jet or sheet before disintegration occurs (prefilming atomizers). In this continuum flow situation, the fluid is often accelerated and the jet or film thickness is thereby decreased. The instability that finally leads to disintegration of the melt is enhanced in this way (see e.g. Gerking, 1993; Schulz, 1996). In metal melt atomization for spray forming application, besides twin-fluid atomization with inert gases, rotary atomizers are sometimes also used (involving centrifugal spray deposition, CSD). Zhao *et al.* (1996a) have described the behaviour of a melt film on a rotating disc with respect to film thickness and velocity distribution. The potential of CSD is limited to ring and tubular deposits and, therefore, suitable only for some specific spray forming applications.

Calculation of two-fluid flows with a continuous phase boundary is possible in several arrangements based on different modelling concepts for the solution of conservation equations. In addition, as essential submodels, suitable methods for transient calculation of the movement and geometry of the phase boundary and inclusion of surface tension effects need to be derived.

Marker and cell method (MAC)

Within the marker and cell (MAC) method, the computational domain where the two-phase flow is to be described is at first divided into spatially fixed discrete elements or volumes. These grid cells reflect the numerical solution of conservation equations. Each grid cell is characterized by its volumetric fluid content (as the ratio of volume of fluid to volume of gas within a cell):

- f (full cell) or $\alpha_f = 1$, for totally fluid filled grid cells;
- b (boundary cell) or $0 < \alpha_f < 1$, for partially filled grid cells on the phase boundary;
- e (empty cell) or $\alpha_f = 0$, for empty grid cells (totally gas filled).

Dependent on the local volumetric fluid content, the conservation equations for mass and momentum of the fluid are numerically solved in the totally filled grid cells ($\alpha_f = 1$). Within the partially filled grid cells, boundary conditions are applied that are based, for example, on vanishing tangential stresses at the phase boundary:

$$\frac{\partial u_t}{\partial x_t} = 0. \tag{4.9}$$

When surface tension effects are neglected, the normal stress at the phase boundary is related to the local pressure only:

$$p = p_u. \tag{4.10}$$

The gas flow field and its exchange with the fluid flow, and also its effect on the phase boundary distribution, are not taken into account within this model (see Reich and Rathjen, 1990).

Transient calculation of the phase boundary geometry, within the MAC method, is based on a simple interface tracking of marker particles. These marker particles are initially positioned continuously on the phase boundary and are tracked as passive tracers whose positions are determined by the calculated velocity distribution within the liquid when the flow field evolves. The local velocity value at the position of the marker particle is derived from neighbouring grid cells by interpolation (see Welch *et al.*, 1966; Hirt *et al.*, 1975). The MAC method, in general, is complicated to handle, as it needs proper adding or rearrangement of the marker particles in strain interface flows, which conventionally needs to be done manually.

Volume of fluid (VOF) method

By far the most popular method of modelling two-fluid situations with continuous phase boundaries is based on a full representation of both fluids in a Euler/Euler model. Here the conservation equations (mass, momentum and thermal energy or mass balances, as well as turbulent kinetic energy and dissipation rate) are derived and solved for both phases separately. In the derivation of this modelling approach, the conservation equations (see Eq. (3.1)) are multiplied by the corresponding volumetric contents α_g and α_f, respectively, of both phases and the additional algebraic constraint is:

$$\alpha_g + \alpha_f = 1. \tag{4.11}$$

Exchange of momentum, thermal energy or mass (transfer rates) is achieved by suitable approaches, which are incorporated as source or sink terms into the conservation equations (Spalding, 1976).

The volume of fluid (VOF) method (see Nichols *et al.* (1980), Torrey *et al.* (1985, 1987)) characterizes the state of the phase at each point in the flow field by introducing the local volume ratio of liquid and gas F as a characteristic scalar function. With respect to its state:

- $F_f = 0$, for empty cells;
- $0 < F_f < 1$, for partially filled cells at the phase boundary; and
- $F_f = 1$, for entirely liquid filled cells.

Conservation of the volumetric gas/liquid ratio and the geometry of the phase boundary are described by the convection equation:

$$\frac{\partial F_f}{\partial t} = -\nabla(\mathbf{u} F_f).$$
(4.12)

Minimization of numerical diffusion effects is necessary in this method and is done, for example, by reconstruction algorithms especially developed for taking surface tension effects into account (Karl, 1996).

Simplification: homogeneous two-phase model

A simplification of the Euler/Euler model for analysis of multiphase flows is based on a homogenization approach. In this approach, a common field distribution (for example, a common velocity distribution in the case of a laminar flow without heat and mass transfer) is defined for both phases. In cases where the coupling and the transfer rates between both phases are important, the velocity and temperature distribution of such a two-phase flow is in thermal equilibrium and can be compared to a single-phase fluid. In this case, only a single set of conservation equations for variable density (gas or fluid) and diffusivity is to be solved. The volumetric fluid content of each cell is calculated separately by two individual continuity equations. The geometry of the phase boundary is derived from the filling function F, as in the VOF method.

Such a homogeneous two-phase model will be used here for analysis of the melt jet prior to atomization. It is a numerically easier to handle alternative than the fully coupled approaches mentioned above for flow situations with two participating fluids and a clearly determined phase boundary.

Mathematical formulation of surface tension effects on the phase boundary may be based on the continuum surface force model (CSF) of Brackbill *et al.* (1992) and Kothe and Mjolsness (1992). This model describes the surface tension force as a continuous three-dimensional effect on the phase boundary, where the filling function has values between 0 and 1. The force on the phase boundary due to surface tension σ is:

$$F_\sigma(\mathbf{x}) = \sigma \kappa_0(\mathbf{x}) \nabla F_f,$$
(4.13)

where $\kappa_0(x, y, z)$ is the local surface curvature depending on the normal vector on the phase boundary and ∇F_f is the gradient of the filling function.

Other modelling approaches

Another modelling approach for continuous two-phase flows is based on the description of fluid dynamics by the Boltzmann equation. Boltzmann models represent the fluids by a number of discrete particles that have their individual position and velocity marked on several grid layers (lattices). Following Shan and Chen's (1994) approach, a two-fluid flow field is described by its potential and interaction of the particles with their individual neighbours through the resulting force:

$$\mathbf{F}(\mathbf{x}, t) = -\frac{G}{\delta t} \psi_f(\mathbf{x}, t) \sum_j \sum_i \psi_f(\mathbf{x} + \mathbf{e}_{ji} \delta t, t) \, \mathbf{e}_{ji}, \tag{4.14}$$

for j grid lines with i velocities at each grid point (x, y, z) with respect to time. The parameter G is a measure of the strength of the interparticle forces and the function $\Psi_f(x, y, z, t)$ is a function that needs to be chosen based on the fluid density with respect to the equation of state of the gas phase. On a small-scale base, the behaviour of the fluid is described by the distribution density of the particles $f_{ji}(x, y, z, t)$. This distribution is derived from the discrete form of the Boltzmann equation as:

$$f_{ji}(\mathbf{x} + \mathbf{e}_{ji}\delta t, t + \delta t) - f_{ji}(\mathbf{x}, t) = \Omega_{ji}, \tag{4.15}$$

for discrete particle velocities at a collision rate Ω_{ji} of particles. Comparison of modelling and simulations of a two-phase flow with a continuous free surface based on Navier–Stokes and Boltzmann equations has been performed, for example, by Schelkle *et al.* (1996). Several sample calculations of droplet flow may be found in the book by Frohn and Roth (2000).

Results for the fluid jet

The undisturbed free flow of a fluid jet from a circular aperture in the direction of gravity (at first, only for small or moderate fluid exit velocities) results in an acceleration of the fluid. The diameter of the jet will decrease with increasing nozzle distance while the mean velocity increases (see principle sketch in Figure 4.8). At some distance from the nozzle, external or internal perturbations may lead to instabilities on the phase boundary, which will result, at first, in a characteristic wave motion of the phase boundary before disintegration of the fluid jet occurs. Several classical investigations have analysed the characteristic length of such jets before atomization (see e.g. Beretta *et al.*, 1984). For a jet which is finally disintegrated by means of symmetrical surface waves on the phase boundary, the normalized disintegration length is connected to the Weber number We as:

$$\frac{L_c}{d_0} = \ln\left(\frac{d_0}{2\delta}\right) \mathrm{We}^{0.5}. \tag{4.16}$$

The parameter $\ln(d_0/2\delta)$ describes the ratio of the fluid jet diameter d_0 to an initial perturbation wavelength δ of small axial-symmetric perturbations, which needs to be derived from experiments dependent on fluid flow state of the emerging fluid jet from the nozzle.

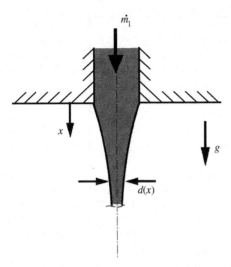

Fig. 4.8 Fluid stream (jet) in gravity field

For a laminar jet one can define (Bayvel and Orzechowski, 1993):

$$\ln\left(\frac{d_0}{2\delta}\right) \sim 24, \tag{4.17}$$

but the ratio depends on the experimental boundary conditions of the flow field and is very sensitive to internal perturbations of the flow field. For a turbulent jet, internal perturbations are not important and the length and geometry of the nozzle where the jet exits are also not important. In this turbulent case, the parameter (Bayvel and Orzechowski, 1993) is:

$$\ln\left(\frac{d_0}{2\delta}\right) \sim 4. \tag{4.18}$$

The resulting disintegration length derived from these correlation equations is an order of magnitude above the free-fall height of the metal melt from the tundish exit to the atomization area for a typical spray forming nozzle.

The result of calculation of fluid jet behaviour by means of a homogeneous twin-fluid model for a cylindrical melt jet of 4 mm diameter (which is a typical scale for spray forming applications) for different Reynolds numbers is illustrated in Figures 4.9 and 4.10. Here, jet diameter versus nozzle distance is shown in Figure 4.9, and the cross-section-averaged melt velocity versus nozzle distance is shown in Figure 4.10. The nozzle distance is normalized by means of the initial melt jet diameter d_0.

The distance between the melt jet exit from the tundish to the point where the melt is atomized by the gas jets for a typical free-fall atomizer within spray forming applications is approximately 100 mm. This is a typical normalized distance of $x/d_0 = 25$. In the case of Re = 10 000, decrease of the melt jet diameter prior to reaching the atomization area is approximately 10% of its initial exit diameter. Based on these results, the following analysis of the disintegration process itself, must take this decrease in melt jet dimension into account before atomization.

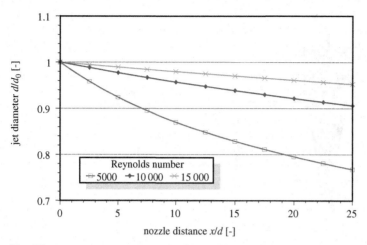

Fig. 4.9 Liquid jet diameter as a function of height

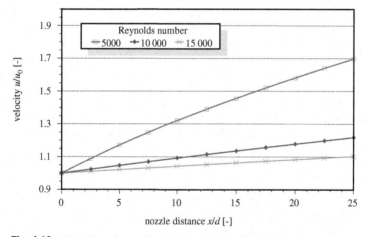

Fig. 4.10 Variation of mean liquid jet velocity with height

In comparison to the two-dimensional results of the numerical model described above, results of a classical one-dimensional analysis based on first principles may be done. Neglecting surface tension and frictional effects on the melt jet behaviour in a first approach, the velocity distribution of the falling jet is:

$$u = \sqrt{u_0^2 + 2gx},$$ (4.19)

and, therefore, the jet diameter can be calculated from continuity as:

$$\frac{d}{d_0} = \left(1 + \frac{2gx}{u_0^2}\right)^{-0.25}.$$ (4.20)

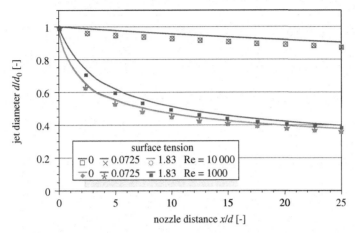

Fig. 4.11 Liquid jet diameter as a function of height for different surface tension values

By adding surface tension effects into this one-dimensional analysis (see Anno, 1997; Schneider and Walzel, 1998; Schröder, 1997), the following fourth-order non-linear equation may be derived describing the contour behaviour of the fluid jet in a gravity field as a function of nozzle distance:

$$\frac{d}{d_0} = \left[1 + \frac{2gx}{u_0^2} + \frac{4\sigma_l(d - d_0)}{\rho u_0^2 d d_0}\right]^{-0.25}. \tag{4.21}$$

Numerical evaluation of this equation is shown in Figure 4.11 for the jet contour of a melt jet (here a steel melt has been taken) for different values of surface tension: $\sigma_1 = 0$, without surface tension effects; $\sigma_1 = 0.0725$, surface tension value of water; and $\sigma_1 = 1.83$, surface tension value of a steel melt. Two different Reynolds numbers are plotted. In spray forming, typically high Reynolds numbers, e.g. Re = 10 000, are achieved at the tundish tip where only negligible influence of the surface tension on the melt jet behaviour for steel melt can be observed. The different graphs in Figure 4.11 are more or less identical. In this one-dimensional analysis, the effect of surface tension slows down the decrease in melt jet diameter. The diameter at a certain distance from the nozzle of the melt jet increases with increasing surface tension values.

If the results of the two-dimensional numerical analysis of the melt jet behaviour are plotted together with the analytical result of the one-dimensional approach on the same diagram, as in Figures 4.10 and 4.11, no visible deviations (actual deviations are below 0.5%) may be observed. This result shows that the influence of surface tension, as well as that of fluid viscosity, on melt jet behaviour in this application (in terms of the jet contour) may be neglected in the first instance. The radial velocity distribution in the melt jet is equally distributed; no significant velocity profile is found.

The aforementioned results have been derived for a stagnant gas surrounding the melt jet. Figure 4.12 shows the result of a homogeneous simulation, where the melt jet is surrounded by coaxial gas flow at different velocity levels (from 0 to 100 m/s), and for a single melt Reynolds number of Re = 10 000. The velocity value on the melt jet centre-line is plotted.

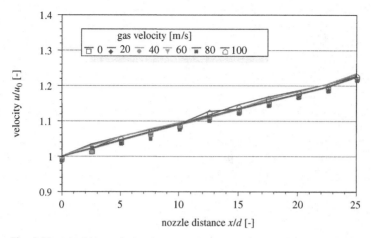

Fig. 4.12   Liquid jet velocity for coaxial gas flow (Re $= 10\,000$)

The coaxial gas flow field has only a slight influence on the melt jet behaviour prior to atomization. The centre-line velocity increases only slightly with increasing gas velocities. In the numerical calculation already presented, jet oscillations at large wavelengths and fluctuations of the melt velocity can be found at increasing gas velocities, which must be regarded as numerical instabilities of the model.

As shown in the discussion on exit profiles of the melt jet from the tundish model, a more or less developed fluid velocity profile may be found at the tundish exit. In the case of a fully developed parabolic (laminar) velocity profile of the melt jet at the tundish exit, Middleman (1995) derived an approximate solution of the decreasing jet diameter (neglecting surface tension effects) as:

$$\frac{d}{d_0} = \left(\frac{4}{3} + \frac{2gx}{u_0^2}\right)^{-0.25}. \tag{4.22}$$

A numerical calculation of an exiting fluid jet having a velocity profile, taking into account friction, gravity and surface tension, has been performed by Duda and Vrentas (1967). In this case, boundary layer approximations of the momentum conservation equations in boundary fitted coordinates have been done for Re > 200. Their results indicate that flattening of the velocity profile developed in the jet due to low tangential stresses on the jet surface is very slow.

The result of the numerical calculation of this problem, with a velocity profile developed in the initial jet at the tundish exit is shown in Figure 4.13. The velocity distribution is shown in the melt jet as well as in the surrounding gas phase, which is initially at rest. The melt jet diameter at the tundish exit is $d_0 = 4$ mm. The velocity at the phase boundary between the melt and the gas increases much faster than in the inner area of the melt flow. In summary, a melt jet diameter with initial velocity distribution, decreases slower than that with constant initial velocity within the melt jet at identical melt flow rates.

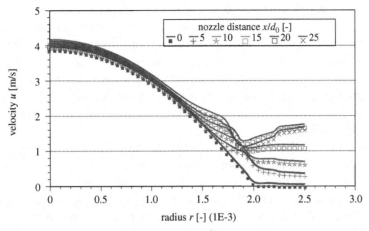

Fig. 4.13 Velocity profile within the melt jet (Re = 10 000)

The use of inline measurement devices during melt atomization offers many technical advantages, such as indirect measurement of the melt flow rate by detection of the melt jet diameter (by optical methods) and melt surface velocity (by laser Doppler anemometry, LDA). These allow derivation of the actual mass flow rate, which is an important process parameter to be controlled. As the above results indicate, not only the (measurable) surface velocity of the melt jet, but the velocity distribution inside the melt jet as well as the development of the jet geometry increasing nozzle distance must be taken into account in order to determine the average mass flow rate.

4.2   The gas flow field near the nozzle

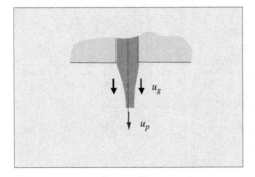

The gas flow field in the vicinity of the atomizer nozzle contributes to melt stream instability, by introducing normal or tangential stresses and momentum transfer, which finally cause disintegration of the melt. In addition, the flow of gas may guide or shield droplets during initial spreading of the resulting spray. Thus, back-splashing droplets (which may result from earlier fragmentation stages of the melt or turbulent dispersion effects) are guided in the main flow direction downwards with the spray. Thereby, nozzle wetting or clogging from recirculating particles is prevented (see below). This shared action of the gas flow field has already been demonstrated in Figure 4.1 in the free-fall atomizer configuration shown. While upper flow of the atomization gas in the primary gas nozzle is responsible for the shielding action, the lower gas stream from the secondary gas nozzle in the atomizer causes the main disintegration process of the melt stream. Both tasks and processes will be analysed later in the chapter.

Due to the angle of inclination of the atomizer gas jets (secondary gas flow) and the open configuration of the free-fall atomizer under investigation here, during atomization localized recirculating gas flow regions may occur. Particles entering these recirculation areas may back-splash towards the atomizer. Thus, metal droplets are transported against the main flow direction of gas, and spray towards the melt stream exit at the tundish tip from below. These particles may stick at the lower edge of the tundish during impact and may cause the growth of a solid layer in this area, resulting in severe process problems. In extreme cases, these metal particles may cause continuous reduction of the free-stream area of the melt flow, and lead to complete clogging of the melt flow and nozzle. In such cases, the entire atomization process comes to an undesired end.

4.2.1   Subsonic flow field

Figure 4.14 shows the main flow directions of the melt and gas (atomizer gas mass flow $\dot{M}_{gg}$, primary gas mass flow $\dot{M}_{g1}$, melt mass flow $\dot{M}_s$) in the vicinity of the atomizer. A photograph of liquid fragmentation in the vicinity of the nozzle during the atomization of water in a free-fall atomizer under certain (critical) operation conditions, is superimposed in the figure. At the theoretical atomization point, the existence and movement of individual droplets

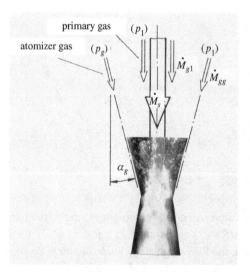

Fig. 4.14   Principle of particle recirculation in the flow field of a free-fall atomizer

above the main disintegration area can be seen. This is the point at which individual inclined gas jets (angle α_g) intersect with the centre-line of the nozzle and melt stream. Obviously, in reality, the disintegration process is initiated well above this theoretical atomization point (as will be discussed below). The droplets shown move within a concentric vortex, upwards in the inner region close to the melt stream surface and downwards in the outer vortex region in the area of the downward pointing jet flow.

This kind of particle recirculation is well known within free-fall atomizer applications and within other twin-fluid atomizer configurations under certain operating conditions. In order to investigate the conditions of particle recirculation and, in particular, the operational conditions necessary to prevent recirculation, two kinds of models are presented:

(1) atomization in a water/air system, and
(2) numerical simulation of a pure gas flow in the vicinity of the atomizer nozzle.

Both investigations use geometric nozzle configurations identical to those used for melt atomization during spray forming. Qualitative results are extracted from photographic evidence of water atomization and from experiments based on injected tracer-particle movements within the gas flow field of the atomizer. Here the pure gas flow field is observed without any liquid flow and liquid atomization (i.e. a gas-only flow).

For numerical modelling of the gas flow field in the vicinity of the atomizer, the discrete arrangement of individual gas jets on a circumference around the melt stream is replaced with a slit nozzle configuration having one circular slit for the primary gas and another circular slit for the atomizer (secondary) gas. The total exit area of both gas flow streams is identical to those of the sum of the respective individual gas jets in reality. By this geometric approximation, the computational power needed for the simulation is drastically reduced, as in the slit configuration one can assume circumferential symmetry of the gas flow and thereby reduce the problem from a three-dimensional to a two-dimensional configuration. The difference between two- and three-dimensional atomizer gas flow fields will be discussed later.

Based on the aforementioned assumption, the description of the gas flow field in the vicinity of the atomizer is based on the stationary (in a first attempt) subsonic and turbulent approach described by the compressible conservation equation, Eq. (3.1), in its non-transient (steady-state) form. Boundary conditions for the state of the gas at the nozzle exit need to be described in terms of gas prepressure and exit Mach number, Ma. In the region of small exit Mach numbers (Ma < 0.3), the flow of gas can be assumed to be incompressible and the exiting mass flow rate can be derived from:

$$\dot{M}_g = \rho_g u_a A. \tag{4.23}$$

In the area of compressible, but subsonic, gas flow (0.3 < Ma < 1.0), the mass flow rate exiting from the atomizer is:

$$\dot{M}_g = \mu_f A p_0 \sqrt{ \frac{2\kappa}{(\kappa - 1)RT_0} \left[\left(\frac{p_u}{p_0} \right)^{\frac{2}{\kappa}} - \left(\frac{p_u}{p_0} \right)^{\frac{\kappa+1}{\kappa}} \right] }, \tag{4.24}$$

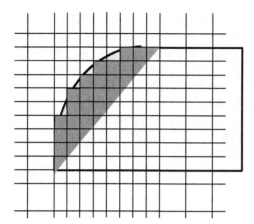

Fig. 4.15 Stepwise approximation of curved boundaries on a Cartesian grid (principle)

and for supersonic gas flow it can be described as:

$$\dot{M}_g = 0.484\ \mu_f A p_0 \sqrt{\frac{2}{RT}} = \text{const.}\ A\sqrt{2 p_0 \rho_0}. \tag{4.25}$$

Supersonic gas flow is discussed in Section 4.2.2. This type of flow is achieved at higher gas prepressures (here $p_0 > 1.89$ bar absolute), where the critical ratio of ambient pressure p_u to atomizer gas pressure p_0 for a two-atomic gas (like nitrogen) is:

$$\left(\frac{p_u}{p_0}\right)_{\text{crit}} = \left(\frac{2}{\kappa + 1}\right)^{\frac{\kappa}{\kappa - 1}}. \tag{4.26}$$

Turbulence modelling for gas flow simulation in this approach is based on the standard k–ε model of Launder and Spalding (1974). Isentropic change of the gas state is assumed. Solution of the resulting system of elliptic partial differential equations is done, in this case, by means of a commercial software package based on a finite volume approach (Rosten and Spalding, 1987). Adaptation of the grid structure to the atomizer geometry to reflect the solid body in the flow field in detail is based on a conventional Cartesian coordinate and grid system. Here the solid surface areas are handled by blocking certain grid cells and approximation of the momentum fluxes on the boundaries of these blocked cells based on no-slip conditions on solid surfaces. Stepwise blocking can be seen in Figure 4.15, approximating curved surface contours. For illustrative purposes only, the grid is shown much coarser here than when it is used in the simulation.

Before using the simulation model to describe the flow field of the atomizer gas, the model should be validated by reference to published experimental results. Similarly, this may be done for geometric models, with simpler flow boundary conditions. For atomizer gas flow, a turbulence-free round jet can be used for comparison; but in this instance, a free coaxial jet is used, where the gas exits perpendicularly from a circular slit in the wall. An example of the simulation of a coaxial jet is illustrated in Figure 4.16, showing a vector of gas velocities in the vicinity of the nozzle. Due to recirculation and the low-pressure region in the inner area of the coaxial jet, the jet contracts despite the initial

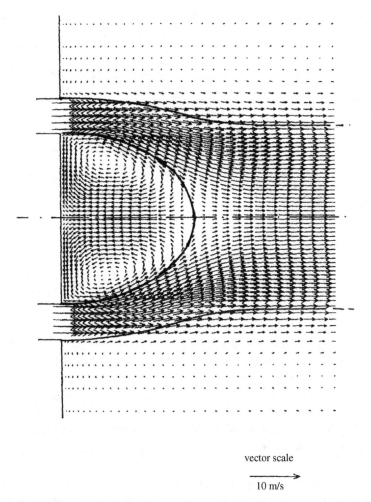

vector scale

10 m/s

Fig. 4.16 Velocity vector diagram of a coaxial gas jet (Fritsching and Bauckhage, 1992)

perpendicular flow direction away from the wall. Some distance from the gas exit, the coaxial jet closes and behaves (after a certain transition region) like a conventional circular turbulent-free jet.

For comparison of simulation results with published experimental results, Figure 4.17 shows the behaviour of axial velocity on the centre-line of a coaxial jet, together with the measured results of Durao (1976). Contraction of the jet, as well as jet spreading at greater distances from the nozzle, are overpredicted by the model, as can be seen by the earlier decrease in gas velocity during simulation compared to experimental values close to the jet nozzle and the slight increase in gas velocity at greater distances. In principle, this unsatisfying result is in agreement with previous studies and with published results of simulation with turbulent-free jets, especially in axisymmetric jets based on the standard $k-\varepsilon$ model (Leschziner and Rodi, 1981; Malin, 1987), where it has been shown that the spreading behaviour of round jets is underpredicted by the simulation model. In addition, for

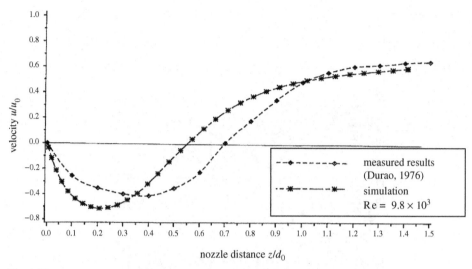

Fig. 4.17   Velocity at the centre-line of a coaxial gas jet (Fritsching and Bauckhage, 1992)

the case of a coaxial jet, contraction of the jet in the vicinity of the nozzle is overpredicted, as can be seen from Figure 4.17 where the free stagnation point is found to be much closer to the nozzle in the simulated rather than experimental results. Based on this evidence, deviations of the same order of magnitude may be expected in the prediction of atomizer gas flow fields. In principle, these deviations in atomization behaviour are unacceptable for proper analysis of the system. However, the principal physical parameters of the gas flow field boundary conditions can still be determined accurately, as these deviations occur steadily where, for example, measures to prevent recirculation can still be derived from simulation results.

Figure 4.18 compares a simulated vector plot of gas velocities in the vicinity of the atomizer for a free-fall configuration with a picture of the actual atomization of water in such a nozzle. The absolute atomization gas prepressure of $p_2 = 1.89$ bar used in this case, causes critical sonic outflow of the atomization gas (here nitrogen) at the gas exit, where the gas exit velocity equals the local velocity of sound at Ma $= 1$. In this first example, the flow field is calculated (and in the corresponding experiment has been used too) without any primary gas application, i.e. $p_1 = 0$ and $\dot{M}_{g1} = 0$. The atomization gas exits the nozzle body at an inclination of $10°$. It can be seen from the figure that the atomized gas stream contracts more intensely than indicated by that inclination angle. Here the same effect, as has been observed for a simple coaxial jet, occurs; the gas stream contracts towards the centre-line of the jet. The point where the gas hits the melt stream is somewhat closer to the atomizer than indicated by the geometric point of jet impingement (theoretical atomization point). Due to gas entrainment from the edge of the circular jet, external gas is accelerated into the inner flow area. Most of the entrainment gas flows through the gap between the atomizer gas ring and the main body of the nozzle. In this gap, a maximum gas velocity of approximately 10 m/s is achieved. In the inner region of the gas flow field, a huge recirculation area can

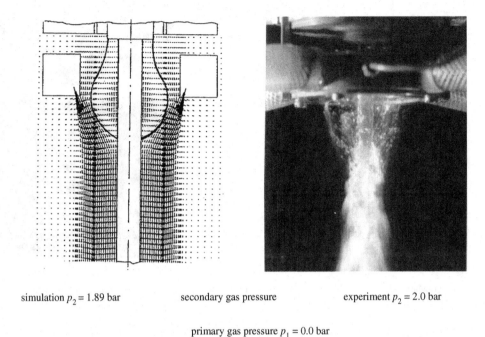

simulation $p_2 = 1.89$ bar   secondary gas pressure   experiment $p_2 = 2.0$ bar

primary gas pressure $p_1 = 0.0$ bar

Fig. 4.18 Comparison of simulated and experimental results for $p_1 = 0$ bar (Fritsching and Bauckhage, 1992)

be observed. A solid line within the vector plot marks the stream-line boundary of this recirculation area. Within the vortex the maximum velocity of the upward-directed gas is 45 m/s. Individual particles entering this vortex, will be accelerated upwards in the direction of melt exit within the main body of the atomizer. These recirculating particles may hit the main body of the nozzle close to the melt exit. This result is validated by visual observation of water atomization under almost identical process conditions as in the simulation shown in the right-hand side of Figure 4.18. Individual droplets can be seen above the atomization area in this figure, which are transported upwards against the main gas flow direction and towards the main body of the atomizer and tundish.

When adding a primary gas flow to this configuration at a low gas prepressure of only $p_1 = 1.1$ bar absolute (0.1 bar above ambient pressure), the gas flow field illustrated in Figure 4.19 results. The primary gas flow that exits the main body of the atomizer close to the melt exit is also in the form of a coaxial jet, which contracts in the manner indicated above. Therefore, in the inner area of the primary gas jet, first a recirculation area can be seen where, due to jet contraction, the flow of gas is deflected radially inwards towards the melt stream, and, afterwards, flows for a certain distance when observed from the melt exit, the gas flows tangentially to the melt jet. With increasing distance from the atomizer, the primary flow of gas is influenced by the much faster rate of flow of the atomizer (secondary) gas. Due to this influence the initial flow of gas again detaches from the central melt stream and is sucked into the faster atomization gas flow. By doing so, a second recirculation area

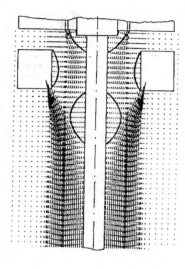

simulation $p_2 = 1.89$ bar   secondary gas pressure   experiment $p_2 = 2.0$ bar

primary gas pressure $p_1 = 1.1$ bar

Fig. 4.19 Comparison of simulated and experimental effects for $p_1 = 1.1$ bar (Fritsching and Bauck-hage, 1992)

within the gas flow field occurs from the point of detachment of the primary gas to the point of impingement of the atomization gas. A third area of recirculating gas can be seen in this simulation at the inside of the atomizer gas ring. This detachment is due to the sharp upper edge of the atomizer gas ring where entrainment through the gap releases the gas from the atomizer ring, while allowing flow around the corner. For comparison with simulation results, the actual tracer flow trajectories through a gas flow field are shown on the right-hand side of Figure 4.19. In this photograph, sand particles have been introduced into the flow of gas at a certain position using a hollow cylinder, i.e. the particles are introduced into the centre of the configuration (Uhlenwinkel *et al.*, 1990). After entering the flow field, these tracer particles are immediately accelerated upwards due to the rising flow of gas in the inner part of the main vortex. Within a few centimetres, the flow direction of these tracer particles is reversed. At this point, the gas velocity is directed downwards, the tracers are transported somewhat radially outwards, and these tracers are then accelerated downwards in the direction of flow of the main atomization gas. The behaviour of these tracers validates the simulation results of a locally bounded recirculation area within the gas flow field for these specific operational conditions of the atomizer. This recirculation zone is surrounded by the main atomizer gas, which is directed downwards.

By increasing the primary gas pressure further, its detachment point is pushed downwards and, finally, the lower recirculation area vanishes and the entire flow in the atomization-relevant flow area is directed downwards. Due to the increased kinetic energy of the primary gas, this part of the flow stays attached to the central melt stream without being sucked into the atomization process. Both of the other vortexes in the flow field remain, but the velocity

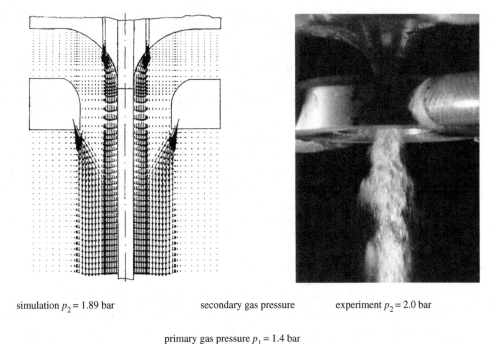

simulation $p_2 = 1.89$ bar   secondary gas pressure   experiment $p_2 = 2.0$ bar

primary gas pressure $p_1 = 1.4$ bar

Fig. 4.20 Comparison of simulated and experimental results for $p_1 = 1.4$ bar (Fritsching and Bauckhage, 1992)

in the upper vortex, close to the melt flow exit, increases strongly (Fritsching and Bauckhage, 1992).

Based on the above conditions for varying gas mass flow rates and prepressures, the optimum conditions for nozzle configuration and geometry may be derived. The main optimization criteria are:

- the construction of a smooth nozzle contour that follows the natural streamline curvature of the flow field to prevent detachment of flow at edges of the atomizer,
- placement of the primary gas exit (gas nozzle) as close as possible to the melt stream exit to suppress recirculation of the primary gas flow in this area,
- increase in the distance between atomizer gas ring and the main body of the atomizer in order to allow maximum gas entrainment through this gap from the spray chamber.

The simulated and experimental effects of an optimized nozzle contour design based on the above criteria is illustrated in Figure 4.20. Here, the operation conditions are chosen for a primary gas pressure of $p_1 = 1.4$ bar absolute at an atomization gas pressure of $p_2 = 1.89$ bar (critical gas flow exit condition). The simulation result shows a gas flow field without any recirculating vortex in the vicinity of the nozzle. But from the right-hand side of Figure 4.20, it is apparent that primary excitation of the atomized fluid is increased because of the proximity of the gas nozzles to the liquid jet, and thus the gas velocity values in the boundary layer between the fluid/gas interface are increased. Here again, water is used as the atomization fluid for the optimized nozzle configuration shown, and only that part of the flow above of the main atomization area is to be seen.

In summary, in order to achieve a recirculation-free gas flow field in the vicinity of a free-fall atomizer, application of a minimum amount of primary gas is needed. The prepressure and flow configuration of the gas must fit the operational parameters of the nozzle tip and are determined by the gas pressure during atomization and by the actual geometry of the atomizer.

This analysis of the subsonic flow field has shown the importance of the primary gas in achieving a stable and recirculation-free situation. The principal interaction of primary and secondary gas flows has been clarified. In the next section, supersonic flow of atomizer gas jets in the vicinity of the atomizer will be discussed, as this type of behaviour is more relevant in practice. Specifically, the role of the atomization gas in achieving optimum melt disintegration will be derived and discussed.

4.2.2   Supersonic flow

In most spray forming applications, the gas prepressures used are beyond the critical pressure level, and are in the range of approximately 3 to 8 bar absolute. Therefore, depending on the contour of the gas nozzle, the flow in the vicinity of the nozzle may exceed the sonic velocity, achieving supersonic conditions. In this range, additional flow field features, such as expansion and compression zones (shocks), occur. Shock structures do not feature in the simulation model discussed in the previous section. But it is important to account for compressibility effects, as the flow of gas from the secondary gas nozzle mainly determines the disintegration efficiency and atomization behaviour of the atomizer system.

One aim of investigating the gas flow field in the vicinity of the atomizer at relevant operational conditions is the description of gas velocity and pressure distribution (including expansion and compression zones) for various atomizer geometries and operational conditions for secondary gas flow. In connection with this aim, the interaction of individual gas jets in a discrete gas jet arrangement (two-dimensional versus three-dimensional behaviour) should be described in terms of its impact on the resulting gas flow situation in the atomization zone.

By modelling gas flow behaviour and investigating the main flow field in the atomizer, the optimum flow conditions may be determined, which may be aimed at specific spray forming operations:

(1) minimization of total gas consumption during atomization and thereby minimization of operational costs throughout the process,
(2) shifting and/or influencing of the resulting droplet-size distribution in the spray (mean and width),
(3) increasing process safety with respect to such problems as particle recirculation and nozzle clogging,
(4) manipulation and variation of the resulting mass flux of droplets in the spray for flexible variation of the product geometry within spray forming.

Point 1 is of minor importance for industrial applications, as the cost of gas consumption is low in relation to other process costs (based on the use of nitrogen as an atomization gas; for other gases, like argon, this needs to be reviewed). Points 2 to 4 are therefore the main focus of optimization efforts within industrial spray forming applications. Point 3, especially, is important for safe operation of the spray forming process. As discussed previously, during atomization individual melt droplets may be back-splashed towards the atomizer instead of being transported downwards in the direction of the spray. From such back-splashing particles, process problems will arise as these materials will influence the free-fall behaviour of the melt stream. Prevention of droplet back-splashing is, therefore, a very important feature of atomizer construction and the choice of suitable atomizer process conditions. The prevention of gas recirculation in the nozzle hinders the back-splashing of droplets, and is achieved through the selection of a suitably low mass flow rate through the primary nozzle (see Section 4.2.1). In summary, Points 2 and 4 are the main optimization potentials to achieve. These are determined by the behaviour of the spray during separation, and are a direct reflection of the construction and arrangement of the atomizer.

The compressible gas flow field in the vicinity of the nozzle is described by comparison with free gas jets from unbounded and bounded flow configurations. Next, the interaction between individual jets needs to be derived, as the flow of gas in an external mixing atomizer is mainly produced by a discrete jet arrangement.

Principle of underexpanded jets

The flow through conventional straight bore holes or apertures during expansion of gas from an atomizer nozzle at super critical pressure is that of an underexpanded jet. This configuration is described by the remaining overpressure of the gas while exiting from the nozzle. Thereby, the gas is not fully expanded in the exit area and expands in front of the nozzle. The main parameters affecting this behaviour are the driving pressure ratio and the geometric arrangement of the nozzle. Up to a pressure ratio of $p_0/p_u = 1.89$ (for air or nitrogen as atomizer gas, where p_0 is the stagnation pressure in the plenum and p_u is the ambient pressure), the pressure in a straight or simple converging nozzle is monotonically lowered until the ambient pressure is reached. If the pressure ratio exceeds the critical pressure ratio, further expansion outside of the nozzle occurs, and the gas pressure in the exit is above ambient. The gas exits the nozzle in this case at a Mach number Ma $= 1$. The remaining pressure potentials in the gas are decomposed by expansion waves that start at the edges of the nozzle exit (so called Prandtl–Meyer expansion fans). Such expansion waves are reflected from the free-jet boundaries as compression waves (see the theoretical sketch in Figure 4.21) and combine in front of the nozzle to form oblique or straight shocks (at higher pressure ratios).

Such shock waves cause energy losses in relation to their intensity. At even higher pressure ratios, a Mach disc (straight shock front in the centre of the jet) with specifically high losses is formed. Such highly underexpanded flows will form at pressure ratios $p_0/p_u > 3.85$ (see Figure 4.22). Behind the Mach disc the local flow velocity is subsonic. For moderately underexpanded jets at pressure ratios $1.89 < p_0/p_u < 3.85$, the compression waves from

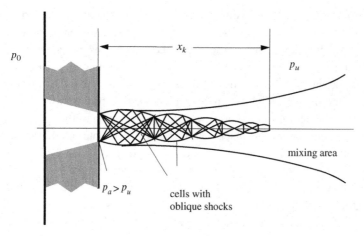

Fig. 4.21   Theoretical sketch of a moderately underexpanded jet

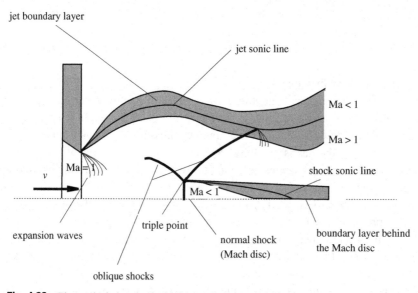

Fig. 4.22   Theoretical sketch of a highly underexpanded jet

the jet boundary coalesce to form oblique shocks. Behind the oblique shocks, supersonic flow conditions still exist (Ma > 1).

Figure 4.21 shows a sketch of a moderately underexpanded single jet flow exiting from a straight or converging nozzle. Inside the shock 'diamonds', strong fluctuation of all flow and gas properties occurs. For an ideal gas, density is proportional to pressure and, therefore, in expansion zones, the density decreases. From continuity, the gas velocity in these areas must increase. After equilibration to ambient pressure ($p = p_u$), which is marked by a supersonic jet core length of x_k, the velocity of the gas further downstream decreases monotonically and its behaviour can be described as that of a turbulent free jet.

The principal flow field of a highly underexpanded jet flow is shown in Figure 4.22. The expansion fan is illustrated, at the edges of the nozzle. The expansion waves are reflected at free jet boundaries as compression waves and coalesce at the triple point to a common straight shock front, which is called the Mach disc (in the case of round jets) or Mach shock. Behind the Mach disc, an area of subsonic velocities exists (Ma < 1); while behind the oblique shocks, the flow field is still supersonic. Therefore, in highly underexpanded jets, even in the area close to the nozzle exit, subsonic flow zones exist, which are covered by supersonic flow areas. The boundaries between supersonic and subsonic flow areas in this region are called sonic lines. These sonic lines, as well as the outer boundaries of the jet, are strongly sheared boundary layer flows, where turbulence is primarily produced. The inner boundary layers inside the shock cell structure, have a high impact on the overall flow field and spreading behaviour of highly underexpanded jet flows, as discussed by Dash and Wolf (1984). Therefore, for modelling and simulation, frictional effects in the vicinity of the nozzle must be taken into account even, or especially, for highly underexpanded gas jets. The set of conservation equations and the solution algorithm for this case must reflect this specific task.

Principle of underexpanded interacting gas jet systems

Within a typical spray forming atomizer a set or system of individual gas jets is located on a common circumferential diameter around the melt flow or atomizer centre-line. Therefore, individual flow and shock structures from neighbouring jet systems may interact and the common shock and flow structure may change. Figure 4.23 shows, as an example, the shock scheme resulting from two interacting jets. If the distance or spacing between the individual jets is decreased, the shock structures may merge to form a single combined Mach disc, as has been shown by Sizov (1991). Also the strength of jet interaction is increased when the pressure is increased, as is the intensity of individual compression and expansion waves.

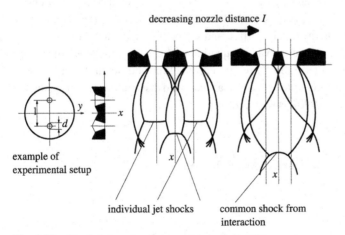

Fig. 4.23   Shock interaction between two adjacent jets

Masuda and Moriyama (1994) have found that during the interaction of neighbouring gas jets: (1) the Mach disc shifts downstream, and (2) the diameter of the common Mach disc is lower than that of individual Mach discs from each single jet. Reduction of the Mach disc size thereby decreases losses by that shock.

Numerical simulation of underexpanded jets

Numerical simulation of underexpanded jet flow is also based on the conservation equations for mass, momentum and energy (Eq. (3.1)). Because of the compressible nature of the gas flow field in this case, an additional relation for density is needed, i.e. (at moderate gas prepressures) the ideal gas law.

Of special interest in modelling gas jet behaviour is the momentum balance; this will enable determination of which kind of forces are mainly responsible for the spreading behaviour of the jet. In the area close to the nozzle, exchange between pressure and momentum forces predominates. Therefore, modelling and simulation in this area is based on inviscid Euler equations without frictional effects (only pressure and inertia). A modified Euler solution code has been used, for example, by Ahrens (1995). Based on a flux-splitting algorithm, the conservation equations are solved on a structured orthogonal grid system. Numerical resolution of the high gradients in shock areas is performed by means of a shock capturing method. By comparing the results of this simulation with experimental data, the important role of viscous forces in the case of underexpanded jets, becomes obvious. Within a few shock cells, a boundary layer at the limits of the jet occurs, which cannot be accounted for by an inviscid code, but is important for the overall behaviour of the underexpanded jet. Therefore, for a closed solution to the spreading behaviour of underexpanded jets also within the far field, a suitable turbulence model has to be incorporated that takes into account the influence of viscous and Reynolds' stresses. Such an extended model has been used by Heck (1998).

In this book, analysis of a viscous, turbulent and compressible gas flow within free jets and atomizers is based on the standard k–ε model (Launder and Spalding, 1974) or, alternatively, on the renormalization group (RNG) model of Yakhot and Orszag. In both models, additional conservation equations for the turbulent kinetic energy (k) and the dissipation rate of turbulent kinetic energy (ε) are derived and solved (two-equation models). For compressible flow simulations, the conservation equation for turbulent kinetic energy is extended by a modified dissipation term. By taking into account the additional dissipation of turbulent kinetic energy in compressible flow fields an extended dissipation term D_k in the k-equation, as derived by Sarkar (1990), is required:

$$D_k = \rho \varepsilon \left(1 + 2\mathrm{Ma}_t^2\right), \tag{4.27}$$

where the turbulent Mach number Ma_t is defined as:

$$\mathrm{Ma}_t = \sqrt{\frac{k}{c^2}}, \tag{4.28}$$

with c the local speed of sound at the related point.

A major problem with the numerical solution of Navier–Stokes equations for transonic flow fields is the coupling of conservation equations through material properties, especially gas density. By preconditioning of the solution matrix, this problem can be overcome (see Weiss and Smith [wei-95]). This technique is based on transformation of the conservation property into a simple variable, which is only related to these variables. A model conservation equation may be formally derived as:

$$\frac{\partial \mathbf{W}}{\partial t} + \frac{\partial \mathbf{F}}{\partial x} = \frac{\partial \mathbf{G}}{\partial x},$$

(4.29)

where

$$\mathbf{W} = \begin{bmatrix} \rho \\ \rho v_x \\ \rho v_y \\ \rho v_z \\ \rho E \end{bmatrix}, \quad \mathbf{F} = \begin{bmatrix} \rho \mathbf{v} \\ \rho \mathbf{v} v_x + p\hat{i} \\ \rho \mathbf{v} v_y + p\hat{j} \\ \rho \mathbf{v} v_z + p\hat{k} \\ \rho \mathbf{v} E + p\mathbf{v} \end{bmatrix}, \quad \mathbf{G} = \begin{bmatrix} 0 \\ \tau_{xi} \\ \tau_{yi} \\ \tau_{zi} \\ \tau_{ij} v_j + \mathbf{q} \end{bmatrix}.$$

(4.30)

A simple, related variable $\mathbf{Q} = f(p, v_x, v_y, v_z, T)$ is introduced whereby the conservation equation my be written as:

$$\frac{\partial \mathbf{W}}{\partial \mathbf{Q}} \frac{\partial \mathbf{Q}}{\partial t} + \frac{\partial \mathbf{F}}{\partial \mathbf{x}} = \frac{\partial \mathbf{G}}{\partial \mathbf{x}}.$$

(4.31)

The Jacobi matrix $d\mathbf{W}/d\mathbf{Q}$ in this case is:

$$\frac{\partial \mathbf{W}}{\partial \mathbf{Q}} = \begin{bmatrix} \rho_p & 0 & 0 & 0 & \rho_T \\ \rho_p v_x & \rho & 0 & 0 & \rho_T v_x \\ \rho_p v_y & 0 & \rho & 0 & \rho_T v_y \\ \rho_p v_z & 0 & 0 & \rho & \rho_T v_z \\ \rho_p H - 1 & \rho v_x & \rho v_y & \rho v_z & \rho_T H + \rho c_p \end{bmatrix},$$

(4.32)

containing the partial derivatives of the gas density as

$$\rho_p = \left. \frac{\partial \rho}{\partial p} \right|_T, \quad \rho_T = \left. \frac{\partial \rho}{\partial T} \right|_p.$$

(4.33)

By means of this preconditioning, better resolution of the velocity and temperature distribution in viscous transonic flow fields is achieved. Based on Venekateswaren *et al.* (1992), even in inviscid fluids the pressure gradients will be determined more exactly. In addition, the mechanical and thermodynamic properties of the fluid are better resolved and solution of this system of equations is simplified for transonic flow fields. In the case of an uncompressible fluid:

$$\rho = \text{const}, \rightarrow \rho_p = 0, \quad \rho_T = 0;$$

(4.34)

for a weakly compressible fluid:

$$\rho = f(T), \rightarrow \rho_p = 0, \quad \rho_T \neq 0;$$

(4.35)

while for a compressible fluid:

$$\rho = f(p, T), \rightarrow \rho_p \neq 0, \quad \rho_T \neq 0. \tag{4.36}$$

Special treatment of the preconditioned matrix is achieved by the so-called flux-splitting algorithm (Roe, 1986). In this algorithm, the operator F in the modified Navier–Stokes equation, containing the convective transport and pressure terms, is discretized in two separate, direction-dependent terms.

An additional difficulty for numerical modelling of transonic flow fields occurs due to the geometric resolution of shock fronts and their numerical representation. Laterally, shock fronts are comparable in dimension to the free path length of the molecules. Therefore, geometrical resolution of shocks on a fixed-structured grid (which should capture the integral flow domain in the same way) is somewhat problematic. In such cases, an unstructured adaptive grid system is preferred. In this way, the grid will be successively adapted during the numerical iteration process. It will be refined in areas where high gradients occur and shocks may be expected. Suitable program modules (such as Rampant, 1996) may be used, which are specifically developed for simulation of compressible hypersonic or transonic, but turbulent, flow fields.

Numerical simulation of individual jets and jet systems

First, results based on numerical simulation with an Eulerian code (Ahrens, 1995; using inviscid assumptions) will be discussed. Figure 4.24 shows a contour plot of the isolines of the time-averaged gas density (isochors) distribution in the nozzle vicinity at a certain pressure ratio of $p_0/p_u = 5$. Gas from the jet flows from left to right. Based on the symmetry of the problem, in this figure only the upper half of the flow domain is illustrated. The lower edge of the figure marks the centre-line of the jet. As the jet configuration is underexpanded at this prepressure, after emerging from the nozzle the jet expands. A normal shock front in the form of a central Mach disc is to be seen in front of the nozzle, which is located at a

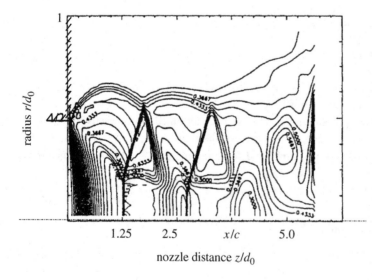

radius r/d_0

1.25   2.5 x/c 5.0

nozzle distance z/d_0

Fig. 4.24 Density distribution of an inviscidly calculated underexpanded jet at $p_0/p_u = 5$ (Ahrens, 1995).

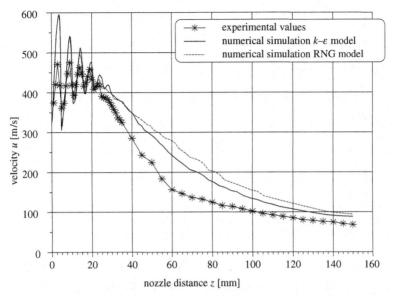

Fig. 4.25 Velocity distribution at the centre-line of an underexpanded jet, comparison between simulation and experiment ($p_0/p_u = 5$) for a single jet configuration (Heck, 1998)

distance of 1.3 nozzle diameters ($x/d = 1.3$) downstream from the exit. Comparison of the location of this first shock front to the experimental results of underexpanded jets of Love *et al.* (1959), shows good agreement (Ahrens, 1995). As the axial distance to the nozzle exit increases, the agreement between simulation and experiment becomes worse, as in this flow area viscous forces become more important but have not been taken into account during the simulation.

Taking into account viscous effects, by solution of the Navier–Stokes equations plus analysis of the turbulence model within the Rampant program, the simulation yields the following results. The velocity distribution at the centre-line of the jet is shown in Figure 4.25, where comparison between simulation and experimental data is achieved by laser Doppler anemometry (LDA). The shock structure is seen in the vicinity of the nozzle. The numerical simulation results are shown to be in good agreement with experimental data: the location and number of shock cells are almost identical. Also the calculated length of the supersonic core of the jet only deviates slightly from experimental findings. Only the decay rate of the gas velocity in the subsonic region is more pronounced in the experiments than in the simulation. Also the amplitudes of the velocity fluctuation differ between experiment and simulation: the peak in velocity values behind the shock front is more intense than has been found experimentally. However, the behaviour of the tracer particles, used for the LDA measurements might cause this experimental deviation. These small, but still inertial, tracer particles cannot follow the steep velocity gradients across a shock exactly.

Comparison of results from simulation and experiments for transonic underexpanded jets shows that, for both approaches, problems may arise which will lead to incorrect results. Some numerical modelling problems that can be applied to this case are now given. A

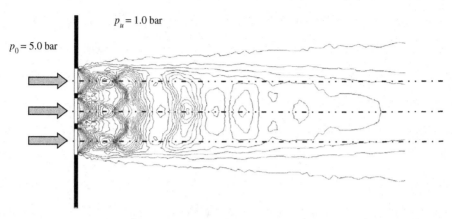

Fig. 4.26 Density contours of an underexpanded jet bundle, three-jet configuration, $d = 2.5$ mm, $t = 0.25$, $p_0/p_u = 5$ (central plane) (Heck, 1998)

pronounced reason for simulation problems is the lag in information regarding initial jet conditions at the nozzle exit. Generally, ideal isentropic behaviour from the plenum into the exit is assumed, characterizing the gas properties in the nozzle exit. Pressure losses due to friction or possible detachment of the jet stream from the wall inside the nozzle are normally neglected. Turbulence properties inside the nozzle are not generally known. Isotropic turbulence models (like the k–ε model) may cause errors in the calculation of free jet flows (as has been previously described). In addition, the calculation for the jet is based on a two-dimensional approach assuming axisymmetric flow. This assumption may be not valid in (underexpanded) jet flows, as transient three-dimensional flow structures, caused by instabilities (Kelvin–Helmholtz instability) at the outer edge of the jet, are important in some regions of the jet. How all these deficits will influence the accuracy of the results of numerical jet calculation is not quite clear and needs further investigation.

As an example of jet interaction, that within a system of three individual circular jets in a row is seen in Figure 4.26. This simulation is done for a fully three-dimensional domain in this case. Plotted are the density contours during the expansion of a three-jet configuration having a spatial spacing of $t/d = 0.25$ at a stagnation pressure of $p_0 = 5$ bar in the plenum. For symmetry reasons, only one-half of the flow field has been modelled. Therefore, only 1.5 jets are reflected in the numerical grid system. The expansion fans are to be seen close to the nozzle exit on the left-hand side of the figure. The compression waves, which are reflected from the jet boundary, combine at the centre-line of the jet system to form a single central shock structure. The coalescence of compression waves from neighbouring jets can be observed. In this three-jet configuration, the shock structure propagates further downstream within the central jet. Here equilibration of pressure occurs more slowly.

Comparison of different jet configurations

During construction of an atomizer system containing a discrete jet configuration, for suitable application of the atomization gas, the geometric configuration and arrangement of the whole jet system (e.g. spacing), as well as the geometry or contour of the individual jets, may

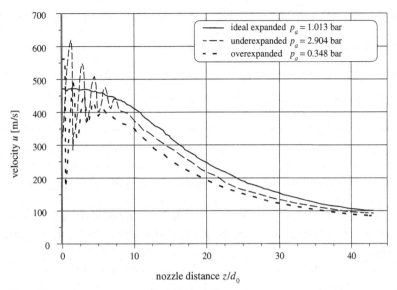

Fig. 4.27 Velocity distribution at the centre-line of under-, ideal-, and overexpanded free jets at $p_0 = 5.5$ bar and $\dot{m} = $ const. (Heck, 1998)

be adapted. Dependent on the individual jet nozzle contour, underexpansion, overexpansion or ideal expansion within a Laval nozzle contour may be achieved. This exit condition has a major influence on the behaviour of the jet. Therefore, overexpanded and ideal expanded jets will now be included in the discussion.

Overexpanded jets, in principle, contain the highest exit momentum at a constant mass flow rate within the three cases mentioned. With the aim of achieving maximum kinetic energy in the atomization area (in order to achieve the maximum slip velocity between gas and fluid) this jet configuration is of special interest. Therefore, the momentum transfer from the nozzle exit to the atomization area where the fluid disintegration occurs needs to be observed.

Comparison of the results of simulation data for overexpanded, underexpanded and ideal expanded gas jets is illustrated in Figure 4.27. The velocities at the jet centre-line for these three cases are plotted versus the nozzle distance for constant prepressures ($p_0 = 5.5$ bar absolute) and mass flow rates. Therefore, the nozzle exit areas of the three jets are different. Their individual values are calculated from the isentropic flow conditions. Therefore, the exit area is highest for the overexpanded jet at identical conditions. Because the overexpanded jet also delivers the highest exit gas velocity, the resulting exit momentum in this case is highest (at constant mass flow rates). The parameters used in the numerical simulation are listed in Table 4.2.

Though the overexpanded jet has the highest exit momentum, this effect starts to vanish in the first shock cell, and one can see that in the transonic region, or at least in the area immediately behind the shock fronts, the underexpanded jet has the highest velocity value. For analysis of the atomization potential of the three individual flow configurations within

Table 4.2 *Numerical simulation parameters for over-, ideal- and underexpanded single gas jets at constant pressure and mass flow rates*

	p_0 [bar]	p_a [bar]	A/A^* [-]	Ma [-]	T/T_0 [-]	v_a [m/s]	I/I_{ideal} [-]
underexp.	5.5	2.904	1	1	0.833	310.5	0.657
ideal exp.	5.5	1.013	1.407	1.77	0.615	472.1	1
overexp.	5.5	0.348	2.57	2.45	0.4544	564.8	1.19

a free-fall atomizer configuration, the behaviour in the far field is especially important. In this subsonic free-jet region, the ideal expanded jet shows the highest velocity values at the centre-line. Though the exit momentum of the overexpanded configuration exhibits the highest value, in the far field it yields the lowest velocity readings of all three configurations. For ideal expanded jet behaviour without any shock structures between the jet exit and the atomization area, the highest potential for atomization purposes is achieved in the far field. Therefore, Laval nozzles are to be used for this configuration in a free-fall atomizer for optimum performance.

Numerical simulation of gas flow near the nozzle of a free-fall atomizer (two-dimensional)

Numerical calculation of the transonic flow of gas in the nozzle of a free-fall atomizer is based on the grid structure illustrated in Figure 4.28. An unstructured adaptive grid has been used: the left-hand side of the figure shows the initial grid, marking the contour of the free-fall atomizer under investigation; while the right-hand side of the figure shows the final grid structure and illustrates local grid refinement in the shock cell zones. Adaptation of the grid structure is based on local pressure gradients. In regions where the pressure gradient exceeds a previously given value, the grid has been refined.

Figure 4.29 shows the calculated flow field, which is based on the grid system afore-mentioned. The numerical calculation is performed on a two-dimensional grid, therefore assuming circumferential symmetry of the atomizer configuration. This means that for simplicity and to keep the numerical effort small, the gas exit has been assumed to have a ring slit configuration where the gas emerges from a ring slit nozzle. The overall nozzle contour is a typical example of a free-fall atomizer as it is used in spray forming applications. The left-hand side of Figure 4.29 shows an underexpanded gas flow and the right-hand side shows the ideal expanded case. Both cases are calculated at constant mass flow rates. In the underexpanded case, the shock cell structure is to be seen in the jet in front of the nozzle: at the listed pressure ratio, no shock occurs at the central plane. The reason is that the ring slit configuration is the extreme case of a discrete jet configuration with zero spacing between the jets. Therefore, the minimum pressure necessary for central plane shock is somewhat above that for a single discrete jet and is not achieved in this example.

The use of ideal expanded gas jets from individual converging/diverging nozzle contours increases the velocity potential in the far field region of the nozzle of interest. Comparison

initial grid   primary gas exit   final grid

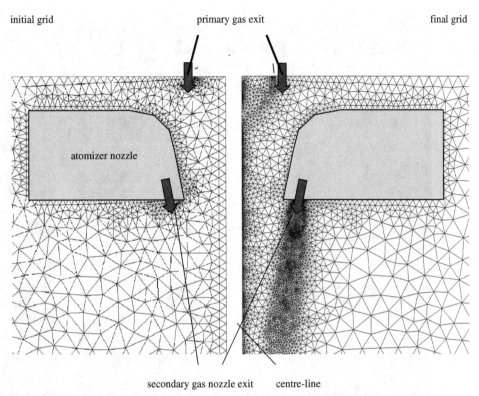

secondary gas nozzle exit   centre-line

Fig. 4.28   Grid structure for the simulation of transonic flow in the atomizer nozzle (Heck, 1998)

of gas velocities at the atomizer centre-line in Figure 4.30 illustrates that in the assumed liquid disintegration area somewhat below the nozzle (approximately 50–100 mm below the nozzle in this configuration) where primary disintegration of the liquid/melt happens, an increase of the gas velocity approximately 10–20% is achieved in this way. In Figure 4.30, the centre of the nozzle ($x = 0$ mm) lies at the lower edge of the separate secondary gas flow nozzle ring.

By using convergent/divergent gas nozzle systems, atomization efficiency is raised and the resulting droplet-size distribution of the process shifts towards smaller diameters (which has been proved experimentally; Heck, 1998). The convergent/divergent nozzle system is more effective than the simple converging one. In addition, the resulting width of the droplet-size distribution during atomization is influenced in this way by changing the individual jet arrangement and geometry. As shown in Figure 4.31, the span of the droplet-size distribution, defined as

$$\text{Span} = \frac{d_{90.3} - d_{10.3}}{d_{50.3}} \tag{4.37}$$

(here measured using water as the atomization model fluid), is lowered, and by using converging/diverging nozzles, the particle size distribution becomes narrower (Fritsching

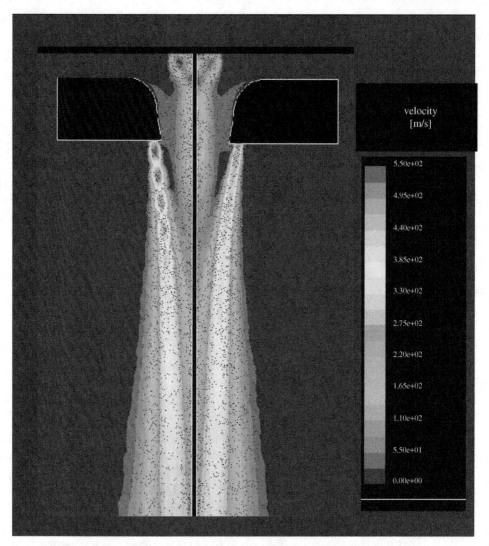

Fig. 4.29 Numerical simulation of the flow field near the nozzle for underexpanded (left side) and ideal expanded (right side) jets. Ring slit nozzle configuration, $p_0/p_u = 5$ (Heck, 1998)

et al., 1999; Heck, 1998). The main reason for this behaviour is the increased production of turbulent kinetic energy within the simple converging nozzle system. As can be seen on the right-hand side of Figure 4.31, where the distribution of turbulent kinetic energy k in the area in front of a converging and a converging/diverging single nozzle is illustrated, within the underexpanded jet issuing from the converging nozzle, in the inner part of the jet additional turbulence due to shear is produced (i.e. in subsonic boundary layer behind the shock front). Therefore, the level of turbulence in an atomizer using discrete simple converging gas nozzles is increased, resulting in broader droplet-size distribution in the spray (increased span).

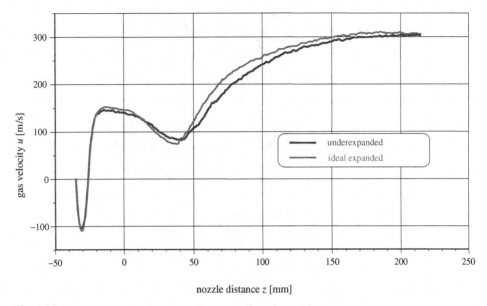

Fig. 4.30 Gas velocity distribution at the centre-line of a nozzle system

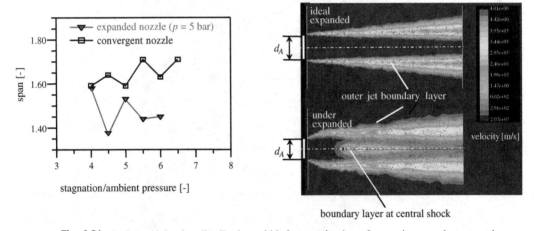

Fig. 4.31 Left: particle-size distribution width for atomization of water in a purely converging gas nozzle system or by means of converging/diverging (expanded) gas nozzles. Right: turbulent kinetic energy distribution in an ideal expanded and underexpanded free jet (Heck, 1998)

Numerical simulation of gas flow near the nozzle of a twin-fluid atomizer (three-dimensional)

The arrangement of gas nozzles in an external mixing twin-fluid atomizer usually corresponds to a number of discrete jets in annular configuration. In a two-dimensional arrangement, the gas exit is described as a slit nozzle configuration. Figure 4.32 simulates the true three-dimensional flow field in an atomizer: this is depicted as velocity isocontour plots at

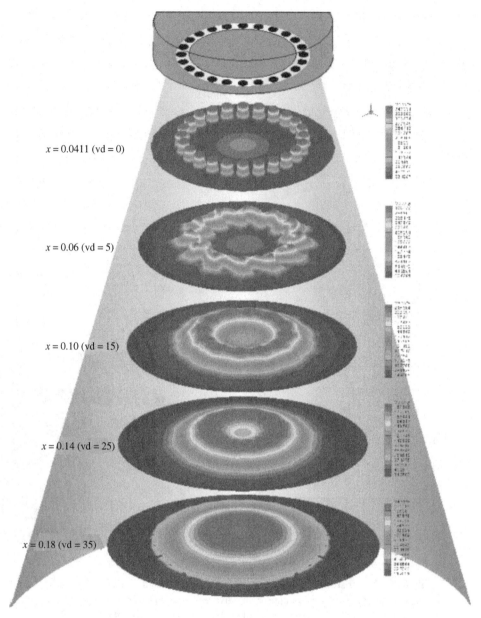

$x = 0.0411 \ (\text{vd} = 0)$

$x = 0.06 \ (\text{vd} = 5)$

$x = 0.10 \ (\text{vd} = 15)$

$x = 0.14 \ (\text{vd} = 25)$

$x = 0.18 \ (\text{vd} = 35)$

Fig. 4.32   Gas flow field in the vicinity of an atomizer: number of jets $= 24$, gas jet diameter $= 4$ mm

distinct distances from the atomizer. The velocity distribution at five different locations is illustrated. The simulation uses the circumferential symmetry of the flow field to calculate the flow as a three-dimensional slice through the atomizer. The plot shows the total flow field, which has been achieved by mirroring the result of the simulation. The colour scaling in each plane is normalized by means of a maximum velocity value in a particular distance from the atomizer. In the upper-most plane $(1/d = 0)$, the 24 individual jets that comprise that particular nozzle arrangement are to be seen at the gas jet exit. In the middle plane at a distance $l/d = 15$ (which is 60 mm below the atomizer in that configuration), the individual jets combine together to form a coaxial jet having maximum velocity at a radial distance outside the centre-line of the configuration. With increasing distance, the local maximum in the velocity distribution moves towards the centre-line of the atomizer. At a distance of $l/d = 35$ (which is 135 mm below the gas jet exit), the maximum velocity value is located almost at the middle of the atomizer configuration and the configuration is comparable to that of a free circular turbulent jet.

In concluding this investigation of the gas flow field near a free-fall atomizer arrangement for spray forming applications, several potential improvements of the configuration can be derived. These optimizations can be achieved by applying suitable process conditions and by suitable configuration of the nozzle, such as:

- improving and adapting the nozzle configuration and regulating the amount of primary gas flowing through the atomizer to prevent back-splashing of melt droplets near the atomizer;
- controlling the kinetic energy during atomization through to increase atomizer efficiency suitable gas nozzle configuration
- configuring the jets to influence droplet-size distribution in the spray, with respect to mean drop size as well as with respect to the width of the drop-size distribution.

The optimum process conditions necessary for construction of a free-fall atomizer, based on those measures aforementioned (i.e. numerical and experimental investigation of atomization models using water), have been validated for molten steel atomization (Heck, 1998; Heck *et al.*, 2000). In these investigations the result of changes in the nozzle arrangement on drop- and particle-size distributions has been analysed, and are in agreement with the results of numerical simulations.

4.3   Jet disintegration

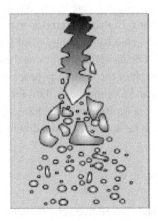

The starting points for all classical investigations of liquid jet disintegration in a gaseous medium are: (1) analysis of the linear stability of the coupled gas/liquid flow field, and

(2) calculation of the capillary instability of cylindrical or plane liquid jets and sheets. Description of fundamental phase boundary behaviour is critical to our understanding of fluid disintegration and atomization, and is key to the development of simple, as well as more sophisticated, atomizer arrangements. The latter involve the production of sprays with monomodal drop-size distribution or single drop production for specific devices, such as ink-jet printers.

Key elements of a liquid disintegration model within atomization are:

- description of the initiation and growth of surface waves on the liquid phase boundary,
- analysis of primary separation of liquid ligaments from the main liquid jet,
- description of successive ligament disintegration into the resulting droplet-size distribution.

A descriptive model for liquid disintegration must provide a general formula for the total initial spectrum of droplet sizes and velocities encountered in the spray, as well as explain the mass flux distribution and spray angle resulting from the fragmentation process. Once these parameters are known, the conditions of the spray at its origin may be derived as a starting requirement for the modelling and simulation of the dispersed spray.

4.3.1   Stability analysis

A cylindrical fluid element in a quiescent gaseous atmosphere is inherently unstable. Small perturbations of the surface (at wavelengths $l > \pi d_l$) will always tend to grow. Initiation of liquid jet disintegration in atomization is described by the initial perturbation and the growth of surface waves at the liquid/gas interface. Instability itself, for example, is due to the establishment of surface tension, leading to a local pressure distribution on the phase boundary dependent on the surface of curvature of the liquid. The first quantitative description of this process was made by Rayleigh (1878), who derived, and solved, the fundamental system of equations describing energy conservation (by neglecting viscous contributions). Rayleigh's analysis is limited to initial sinusoidal perturbations of small amplitude. Weber (1931), in continuation of this work, investigated the viscous system. He neglected inertial effects, and his solution also is only valid for small initial perturbations. Both discovered an exponential growth of small perturbations at unstable wavelengths. Due to the range of process and liquid properties in real atomization processes, viscous effects in the gas flow analysis must be included. The limit up to which an inviscid assumption of the instability process can be made with tolerable error, has been derived by Cousin and Dumouchel (1996) for the case of a planar liquid sheet, and is summarized in terms of a limiting dimensionless number. The influence of viscous forces on jet disintegration can be neglected if the dimensionless number M is:

$$M = \frac{\mu_l \rho_g^2 u_{rel}^3 h_f}{\rho_l \sigma_l^2} < 10^{-3}, \tag{4.38}$$

where the properties of the gas (g) and the liquid (l), and the relative velocity between the gas and liquid u_{rel} and h_f, the liquid film thickness, need to be introduced.

From the pioneering work of Rayleigh and Weber, the following analytical linear stability analyses are all generally limited by one or more basic simplifying assumptions, i.e. neglect of viscosity contributions from either the gas and/or the liquid, partial neglect of inertial effects, or limitation of the analysis to small initial perturbation amplitudes.

Jet instability at small perturbation wavelengths (aerodynamic interaction and liquid disintegration), based on linearized conservation equations and partially neglecting inertial effects and gas viscosity (neglecting the gas boundary layer), was first investigated by Taylor (1940). If, in addition, liquid viscosity is neglected, this results in the classical case of aerodynamic-fluid disintegration based on Kelvin–Helmholtz instability.

A stability analysis that is independent of the ratio of wavelength to jet diameter λ_d/d has been derived by Mayer (1993), based on the contributions of Taylor (1940) and Reitz (1978). The reference system in this analysis is moved with fluid velocity u_l and the results are related to the relative velocity between the gas and liquid $u_{rel} = u_g - u_l$. The complex formulation is based on a perturbation of basic flow in the form of a surface wave with amplitude

$$\eta_S = \text{Real}(\eta_0 e^{ikx+im\theta+\omega t}), \tag{4.39}$$

where $k = 2\pi/\lambda_d$ is the wave number, m is the mode of the perturbation and $\omega = \omega_r + i\omega_i$. The parameter ω is the complex growth rate of the perturbation. The real part of the complex growth rate $\text{Re}(\omega) = \omega_r$ describes the reaction function of the liquid jet by external excitation and the behaviour of the perturbation. For negative growth rates, i.e. $\omega_r < 0$, the perturbation will be damped out; positive growth rates, $\omega_r > 0$, will lead to an exponential growth of the initial wave. In this derivation, the linearized conservation equations for flow of gas and liquid are solved based on the introduction of a stream function ψ, the velocity potential ϕ and the related perturbation formulations. Hereby the velocity of the gas is neglected, resulting in a velocity jump at the phase boundary between liquid and gas. The solution of the equation system is:

$$\omega^2 + 2\frac{\mu_l}{\rho_l}k^2\omega\frac{I_1'(kr_0)}{I_0(kr_0)}\left[1 - \frac{2kl}{k^2+l^2}\frac{I_1(kr_0)}{I_0(kr_0)}\right]$$
$$= \frac{l^2-k^2}{l^2+k^2}\frac{I_1(kr_0)}{I_0(kr_0)}\left\{\frac{\sigma_l k}{\rho_l r_0^2}[1-(kr_0)^2]+\frac{\rho_g}{\rho_l}\left(u_{rel}-\frac{i\omega}{k}\right)\frac{K_0(kr_0)}{K_1(kr_0)}\right\}, \tag{4.40}$$

and contains the modified Bessel functions I_n and K_n of nth order and their derivatives I', as well as $l^2 = k^2 + \omega/v_l$. From the general solution in Eq. (4.40) some specific cases can be derived, which will be discussed in the following.

Large-wavelength area

When aerodynamic effects are neglected, as well as fluid viscosity, gas density and relative velocity between the fluid jet and gas flow field, the classical solution of Rayleigh (1878) is derived:

$$\omega^2 - \frac{\sigma_l}{\rho_l r_0^2}k\left[1-(kr_0)^2\right]\frac{I_1(kr_0)}{I_0(kr_0)} = 0. \tag{4.41}$$

In this equation, I_0 and I_1 are the Bessel functions of zero and first order. As already mentioned, perturbations with wavelengths $\lambda > \pi d_i$ are inherently unstable, resulting always in positive growth rates. If, in the total unstable wave spectrum, only the particular wavelength with maximum growth rate is considered (the fastest growing wave of all unstable waves, representing the maximum instability function), the analysis finally leads to classical Rayleigh disintegration of a liquid jet. This may be achieved by controlled excitation of the liquid jet. In this case, the ligaments from disintegration of the liquid jet form monodispersed spherical droplets at a resulting drop size directly related to the jet diameter by $d_p \sim 1.89 d_i$. This effect is used in monodispersed droplet generators.

Small-wavelength area

In the case of an inviscid flow this is the classical Kelvin–Helmholtz instability derived by Bradley (1973a, b) for stimulation of the primary disintegration process at small wavelengths for coaxial atomization of liquids.

In the limiting case of very small wavelengths ($\lambda \ll d_i$), the solution above leads to:

$$\left(\omega + 2\nu_l k^2\right)^2 + \frac{\sigma}{\rho_l} k^3 - 4\nu_l^2 k^3 \sqrt{k^2 + \frac{\omega}{\nu_l}} + \xi_g (\omega + i u_{\text{rel}} k)^2 \frac{\rho_g}{\rho_l} = 0. \tag{4.42}$$

Introduced in this solution is the assumption that the viscosity of the gas causes instability of the jet, which has been derived by Sterling and Schleicher (1975). These authors found that by accounting for boundary layer effects of the gas flow field, wave growth is stabilized. This can be regarded as a damping effect of the perturbation pressure on the liquid jet surface. The boundary layer coefficient ξ_g is related to this effect, where:

- $\xi_g = 1$, is the case of an inviscid gas flow;
- $\xi_g = 0.8$, takes into account the profile of the turbulent boundary layer of the gas flow field at the liquid/gas phase boundary.

In a similar way, Miles (1957) took into account the turbulent character of the gas flow field at the phase boundary. By neglecting gas viscosity, the logarithmic velocity profile in the fully turbulent region is expressed in terms of a suitable pressure distribution on a wavy liquid surface (Bürger et al., 1989, 1992).

Generally, analyses of aerodynamic disintegration have found that surface tension has a stabilizing effect (shifts maximum unstable wavelength towards lower values) and fluid viscosity a damping effect (lowers the amplitude of the growth rate while keeping the maximum growth rate constant). Increasing the relative velocity between gas and liquid phases always has a destabilizing effect.

The instability of the liquid/gas phase boundary of a jet/sheet is divided into several modes, which can be described by their specific spatial growth functions. Besides the symmetric mode ($m = 0$) of wave growth of phase boundaries, the liquid jet may tend to long oscillatory movement (asymmetric jet oscillations, see Figure 4.33). By neglecting gas and liquid viscosities and the region of high relative velocities between the gas and the

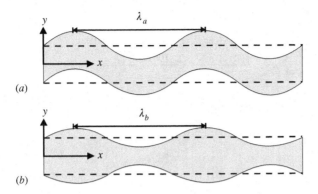

(a)

(b)

Fig. 4.33 Asymmetrical (a) and symmetrical (b) instability modes of liquid jet disintegration

liquid, Levich (1962) analysed the stability of the first asymmetric mode ($m = 1$) of a liquid jet as:

$$\omega^2 + \rho_g k^4 r_0^2 u_{rel}^2 \frac{\left(\ln \frac{kr_0}{2}\right)}{2\rho_l} - \frac{\sigma_l k^2}{2\rho_l r_0} \left(1 - m^2 - k^2 r_0^2\right) = 0. \tag{4.43}$$

In the case of a planar liquid sheet, Hagerty and Shea (1955) have shown that an inviscid fluid has higher growth rate values in the asymmetric mode than in the symmetric mode. Here the asymmetric mode is always dominant.

Despite the relatively crude assumptions and simplifications which have to be made to apply linear stability analyses, the use of this linear theory has a number of advantages in the development of a disintegration model for fluid atomization: first, linear stability analysis is based on simple assumptions and, second, mathematical efforts for its solution are quite low. This makes the handling of this method easy. Linear stability analyses deliver general stability criteria for the liquid/gas system under investigation and allow calculation of wavelengths and growth rates that mainly cover the disintegration process. An overview of classical solutions to the system of ordinary linear differential equations in stability analyses has been given by Markus et al. (2000).

Non-linear stability analyses, taking into account some of those previously necessary simplifications, which have to be performed within linear theory, have been calculated for some geometric boundary conditions and for certain cases. In the sense of stability analyses, the resulting system of conservation equations is solved using a perturbations approach or by numerical analysis with discrete perturbations (see, for example, Dumouchel, 1989; Panchagnula et al., 1998; Rangel and Sirignano, 1991; Shokoohi and Elrod, 1987).

In the following, the results of a linear stability analysis based on the above assumptions for a coaxial twin-fluid atomizer with external mixing and a reference water/air system with a liquid jet diameter of $d_i = 2$ mm will be discussed. Results in terms of growth rates for symmetric ($m = 0$, short wavelength perturbations) and asymmetric cases ($m = 1$, long wavelength perturbations) are shown in Figures 4.34 and 4.35 for variable relative velocity between the coaxial atomizer gas and central liquid. From these results, the principal

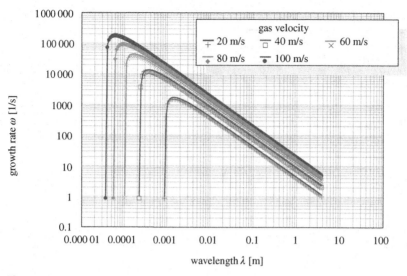

Fig. 4.34 Growth rates of the symmetric mode, $m = 0$, system H_2O/air

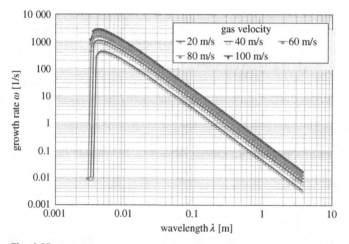

Fig. 4.35 Growth rates of the asymmetric mode, $m = 1$, system H_2O/air

influence of process parameters on phase boundary excitation can be shown. The relative maximum of each growth rate curve illustrates the fastest growing wavelength.

Comparison of symmetric and asymmetric modes in this example shows that symmetric perturbations always yield higher growth rates than asymmetric ones. Therefore, symmetric jet behaviour should finally lead to disintegration in this case. This result has not been confirmed by experimental studies for these boundary conditions so far (Faragó and Chigier, 1992; Hardalupas *et al.*, 1997). For an atomizer of similar type to the one discussed here, Hardalupas *et al.* (1997) found that only a small amount or number of droplets were

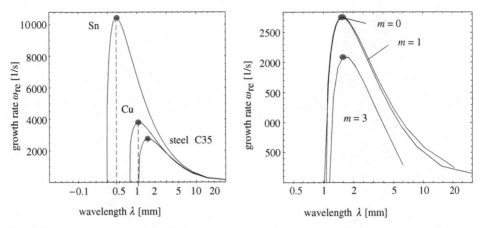

Fig. 4.36 Stability analysis of a liquid metal jet, $u_l = 2$ m/s, $u_g = 100$ m/s, $d_l = 4$ mm (Markus *et al.*, 2000). Left: different melt types. Right: different instability modes for steel melt

immediately sheared off the liquid phase boundary at low wavelengths. These droplets make only a small contribution to the resulting spray; therefore the results of the stability analysis need to be scrutinized or extended to more realistic boundary conditions. In addition, linear evaluation of Figure 4.35 indicates that the wavelengths with maximum growth rate show only a slight dependence on the relative velocity between fluid and gas. This result is, indeed, confirmed by experimental atomization investigations in Hardalupas *et al.* (1997).

Comparison of different modes of stability for a jet of molten metal in a coaxial gas flow has been discussed by Markus *et al.* (2000), based on solution of the stability problem delivered by Li (1995). Results for growth rates of different metal melts are shown on the left-hand side and results for the three basic modes of instability ($m = 0$, 1 and 3) for a steel melt jet are illustrated on the right-hand side of Figure 4.36. The results have been calculated for a jet diameter $d_i = 4$ mm moving at a velocity $v_i = 2$ m/s in a coaxial gas flow at constant velocity $v_g = 100$ m/s. The influence of material properties on melt jet stability is to be seen in the left-hand figure. Here, mainly the effect of the surface tension, increasing from tin ($\sigma = 0.544$ N/m) via copper ($\sigma = 1.31$ N/m) to steel ($\sigma = 1.83$ N/m), can be observed. Increasing values for surface tension result in decreasing Weber numbers, which finally also result in smaller amplitudes of the growth rate. In the same way, the dominant wavelength of jet disintegration increases. In comparison to water jets, for molten metal jets the growth rates are much lower. Here somewhat higher energies (relative gas velocities) are used in order to achieve similar atomization characteristics to those for water. But at higher relative velocities, the significance of higher instability modes increases. Maximum growth rates for the steel melt jet are observed for the symmetric mode $m = 0$, but the asymmetric mode $m = 1$ has almost identical growth rates, while higher modes show maximum growth rates at much smaller values (see the right-hand side of Figure 4.36).

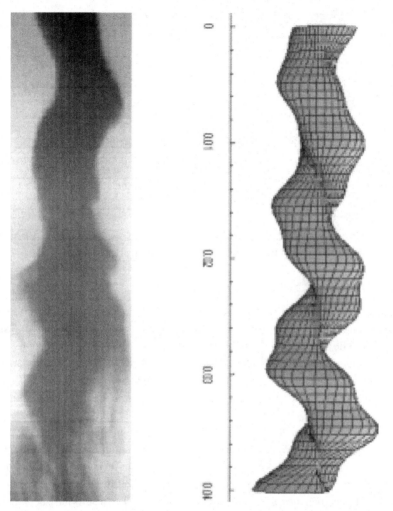

Fig. 4.37 Liquid jet analysis. Left: experiment (steel). Right: theory – superposition of the three basic instability modes (Markus *et al.*, 2000)

A comparison of calculated growth rates and wavelengths from high-speed video pictures for primary perturbation of tin and steel melt jets in Markus *et al.* (2000) have shown that theoretical wavelengths are smaller than those seen experimentally, by one order of magnitude. This has also been found in comparison of experiments, direct numerical simulations and results from linear theory for atomization of liquid sheets by Klein *et al.* (2002). However, superposition of different modes with adapted growth rates (which do not match those in Figure 4.36) results in quite a good picture of the processes involved in primary perturbations of the liquid tin melt jet just in front of the exit from the atomizer nozzle. Figure 4.37 compares the primary excitation of a tin melt jet within a free-fall atomizer nozzle, on the left-hand side, to the three main instability modes on the right-hand side.

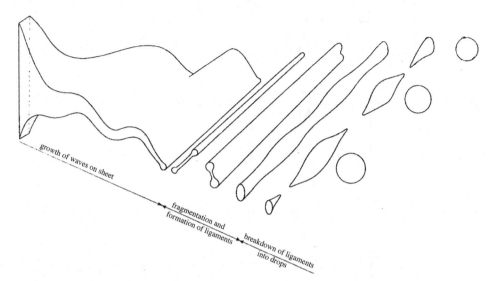

growth of waves on sheet

fragmentation and
formation of ligaments

breakdown of ligaments
into drops

Fig. 4.38   Aerodynamic fluid fragmentation of a plane liquid sheet (Dombrowski and Johns, 1963)

4.3.2   Jet disintegration model

Free-fall atomizer

In the initial models for phase boundary behaviour introduced above, the stability analyses discussed focus on the excitation and initial growth of surface perturbations on the liquid jet. The next step in the disintegration model is to define the point at which the growing liquid element is separated from the main excited jet. Here additional models for liquid separation are needed. First attempts for derivation of a jet disintegration model based on the spray structure and a correlation equation for the resulting droplet-size distribution are to be found in Dombrowski and Johns (1963). Their model for the aerodynamic disintegration of a plane liquid jet emerging from a planar slit nozzle is illustrated in Figure 4.38. Starting from the growth of an instability at the interface, at first plane liquid elements, or ligaments, are separated from the jet in a longitudinal direction, which are deformed to cylindrical elements further downstream. Next, due to capillary instabilities these cylindrical ligaments disintegrate into fragments that build, by the action of surface tension, the resulting droplet structure in the jet. From this initial model, further attempts to describe the continuum disintegration, especially for metal melts within twin-fluid atomization by gases or liquids, are documented in Antipas *et al.* (1993), Bürger *et al.* (1989, 1992) and von Berg *et al.* (1995). The starting point for their models of liquid disintegration is the hypothesis that the unstable wave, having maximum excitation and growth rate ω_r, will prevail against the entire unstable wavelength spectrum and therefore will grow fastest. This wave at the critical wavelength λ_{max} will finally lead to the primary break-up of the liquid jet.

A model for primary liquid disintegration for a cylindrical fluid jet in a coaxial gas flow field will be discussed next, as outlined in Figure 4.39. The primary fragmentation process relating to shear flow instabilities for a cylindrical jet can be described by separation of

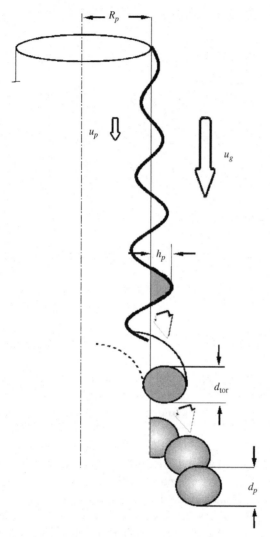

Fig. 4.39 Model of fragmentation and the drop-formation mechanism within coaxial twin fluid atomization

a fluid ligament, which has been formed from sinusoidal surface waves of the liquid jet. The initial ligament volume of this separated liquid element is V_{sin}. This detached liquid element is deformed due to surface tension effects to form a toroidal (ring-shaped) liquid element having identical volume V_{tor}:

$$V_{\text{sin}} = 2(R_l + \pi h_l/8)\lambda_{\text{max}}h_l = 2\pi^2 R_{\text{tor}}^2(R_l + \pi h_l/8) = V_{\text{tor}}. \tag{4.44}$$

This toroidal element is inherently unstable and will fragment, due to capillary instability from surface tension effects, into individual droplets. From the derivation mentioned above

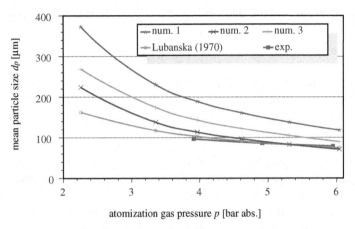

Fig. 4.40 Resulting drop size of the fragmentation model

for high wavelengths at maximum growth rates (Taylor-instability), the resulting droplet-size diameter in the spray can be derived as:

$$d_p = 1.89, \quad d_{tor} = 3.78\sqrt{\frac{\lambda_{max}h_l}{\pi^2}}. \tag{4.45}$$

For evaluation of the separating liquid volume, in addition to description of the critical wavelength λ_{max} with maximum growth rate from linear instability analysis, the critical height of the surface wave at the time of separation from the main jet is necessary. A simple model for the critical ligament height, dependent on the wavelength, has been formulated by Bradley (1973) with the correlation:

$$h_{crit} = \lambda_{max}/4. \tag{4.46}$$

The final fragmentation model for a round molten metal jet in a coaxial gas flow field within a free-fall atomizer consists of the steps:

• analysis of jet instability,
• derivation of maximum unstable wavelength,
• calculation of primary ligament volume (ring),
• derivation of resulting droplet size.

The model determines the mean drop-size diameter of the spray. This model has been used for the analysis of atomization of a steel melt within a free-fall atomizer, and is shown in Figure 4.40. Here the dependency of the mean value of the droplet-size distribution in the spray on atomizer gas pressure p is illustrated.

For stability analysis and derivation of maximum growth rates within the fragmentation model, an estimation of relevant gas velocities has been performed dependent on gas pressure. Here the measurement results of Heck's (1998) investigation have been used for the maximum gas velocities in the atomizer. This maximum gas velocity at the centre-line of the jet, is dependent on gas pressure, and this in turn is dependent on the gas mass flow rate, correlated by:

$$u_{max} = 645.25 \dot{m}_g^{0.585}.$$

(4.47)

The mass flow rate of the atomizer gas is related to the absolute gas pressure in the plenum of the atomizer by:

$$\dot{m}_g = 0.685 \mu_f A \frac{p_0}{\sqrt{RT_0}},$$

(4.48)

which is dependent on the total gas nozzle exit area A. The frictional coefficient for the gas flow emerging from the atomizer has been taken as $\mu_f = 0.8$, in good representation of measurements. From this gas flow description, the principal mass flow rate correlation already described in (4.25) can be derived. Three different relations for droplet sizes in the spray, based on three variations of the critical ligament height prior to stripping from the jet, have been calculated based on the different boundary conditions:

- Number 1: critical ligament height $h_{crit} = \lambda_{max}/2$, boundary layer factor $\xi_g = 1$;
- Number 2: critical ligament height $h_{crit} = \lambda_{max}/4$, boundary layer factor $\xi_g = 1$;
- Number 3: critical ligament height $h_{crit} = \lambda_{max}/4$, boundary layer factor $\xi_g = 0.8$.

Drop-size correlation

For comparison with model results, the data for a representative spray droplet size calculated from Lubanska's (1970) correlation are also plotted in Figure 4.40. This empirical correlation of particle sizes within molten metal atomization is successfully used in a number of investigations for comparison, as it reflects the most relevant influences of physical and process conditions on atomization in terms of median droplet sizes and standard deviations:

$$d_p = d_l K_{Lub} \left[\frac{\nu_l}{\nu_g} \frac{1}{We} \left(1 + \frac{\dot{m}_l}{\dot{m}_g} \right) \right]^{0.5}.$$

(4.49)

In this correlation, the liquid/gas viscosity ratio ν_l/ν_g is used, as well as the liquid Weber number $We = \rho_l u_{max}^2 d_l / \sigma_l$ and the inverse of the mass flow rate ratio (1/GMR; GMR = gas/metal mass flow ratio). The empirical constant K_{Lub} in (4.49) should mainly reflect different atomizer geometries. Lubanska (1970) has given its value to be between 40 and 80, but several other investigations have found much greater values. Several investigations have attempted to verify Lubanska's formula for metal particle-size distribution from melt atomization. Rao and Mehrotra (1980) investigated the influence of nozzle diameter and atomization angle on particle sizes, finding that the mean droplet size decreases with decreasing nozzle diameter and increasing atomization angle. They found a different value for the exponential factor as well as for the atomizer-dependent constant in Lubanska's correlation. Another modification has been proposed by Rai et al. (1985), who studied melt atomization within ultrasonic gas atomizers, and proposed modification of Lubanska's formula.

A discussion and evaluation of the relevance of Lubanska's formula for the atomization of metal melts is to be found in Bauckhage and Fritsching (2000). A broad overview

relative flow

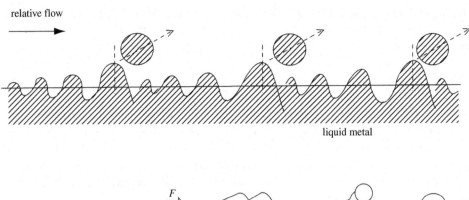

liquid metal

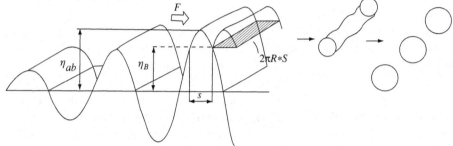

Fig. 4.41   Stripping and drop formation mechanisms (Bürger *et al.*, 1989, 1992)

of a number of empirical correlations for mean and median droplet sizes dependent on operational conditions and nozzle types and geometries for melt atomization within different atomizer configurations can be found in Liu's (2000a) work.

As can be seen from Figure 4.40, taking into account the boundary layer behaviour of the gas at the gas/liquid interface results in larger particle sizes from calculation during atomization. Despite some important quantitative deviations from measurement results, the theoretical approaches described above for a disintegration model obtain useful qualitative assessments of the resulting mean or median droplet sizes during atomization of melts. Also, the effect of changing operational or process parameters and gas or liquid melt properties on the droplet-size spectrum are quite easily assessed by model approaches.

Additional refinements of such fundamental models for primary disintegration within twin-fluid atomization are based on a more detailed analysis of the liquid stripping mechanism from the liquid surface and a description of the most relevant wavelengths at highest growth rates. Here the work of von Berg *et al.* (1995) and Bürger *et al.* (1989; 1992) provides some useful and detailed correlations. In agreement with the earlier investigations of Jeffreys (1924), these authors observed that the gas flow field attacking the wave, under critical conditions, will result in stripping of the wave crest and not of the whole wave volume, while the mean height of the wave remains practically unchanged. A principal sketch of this modelling assumption is shown in Figure 4.41, which also lists the relevant parameters of the model. The turbulent boundary layer character of the gas is taken into account during linear stability analysis of the liquid jet, as in the model of Miles (1957,

1958, 1960, 1961). Starting with an energy balance at the wave crest, where the work done by the gas on the wave crest is related to the free surface energy which is created by the newly generated liquid surface, the critical wave height for stripping $\Delta h = \eta_{ab} - \eta_B$ is:

$$\Delta h \geq \frac{4\sigma_l}{c_d \rho_g u_{rel}^2}. \tag{4.50}$$

In this model a resistance coefficient c_d is introduced for the gas flow at the wave crest. This coefficient takes either a constant value (Bürger *et al.*, 1984) or, alternatively, is derived from the local pressure distribution at the wave form (Bürger *et al.*, 1992). In the latter, the resistance coefficient depends on the local height of the wave. The basic height of the wave η_B is taken as constant with a value of half the wavelength (length to height ratio of the wave is $l/d = 2$). From this value the variable diameter of the stripped liquid elements (in the case of a cylindrical element) d_{cyl} is derived.

In a further refinement of this model, Schatz (1994) linked the primary liquid fragmentation process directly into a simulation of two-phase flow in the disintegration area. Here the behaviour of both phases is simultaneously analysed by conservation equations. First, stepwise stripping of fluid elements from the central jet, as in the above model, is calculated. Then, the resulting decrease of the liquid mass within the remaining liquid jet is derived. The contribution of the stripped fluid ligament to the successive fragmentation process and the resulting drop-size distribution are calculated. Next, development of the now reduced liquid jet diameter is observed and directly related to the local gas flow field at that particular location. The stripped fluid drops and ligaments are introduced as source terms for mass and momentum in the calculation of the multiphase flow field. Continuation of the fragmentation process leads to successive stripping and, therefore, lowering of the jet diameter. The coupling and interaction between the phases is regarded for in this way. The calculated local drop-size contributions of the discrete stripping events are combined to achieve the final droplet-size distribution of the spray. The initial configuration of the spray after atomization is derived in this way. Ongoing further fragmentation of these droplets due to secondary atomization processes is not taken into account. Therefore, the finest droplets in the calculated droplet-size distribution are underrepresented. From these results of model calculation, concepts for optimized nozzle configurations for liquid and melt atomization have been derived (Schatz, 1994).

Close-coupled atomizer

Close-coupled atomizers are particularly common in metal powder production processes. The advantage of this type of atomizer is that it forms a prefilm at its lower edge, which afterwards is disintegrated. The basic concept for modelling the fragmentation process of metal melts within a close-coupled atomizer arrangement has been derived by Antipas *et al.* (1993). This concept deviates from the aforementioned models for free-fall atomizers in terms of the basic fragmentation process and the contribution of the stripped fluid ring (for a cylindrical jet). Instead of using Taylor fragmentation to describe disintegration of the liquid ring, an analysis based on the Kelvin–Helmholtz instability theory is performed for this fluid element. The droplet stripping mechanism begins with the cylindrical shape

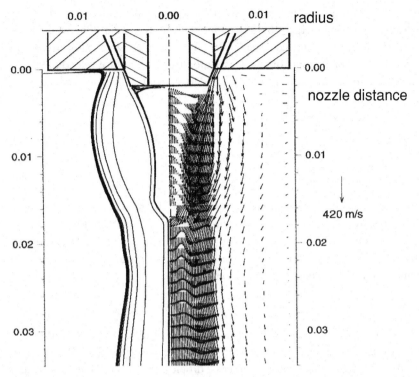

Fig. 4.42 Calculation of the gas flow field in the vicinity of a close-coupled atomizer (Liu, 1997)

of the primary ligament. Not only the fastest growing wave, but the total unstable wave spectrum, contribute to the liquid jet disintegration process (expressed in terms of a lower and an upper bound to the excitation spectrum). The atomizer in this study is operated under supercritical conditions, but the gas velocity used in Antipas *et al.*'s (1993) calculations has been taken as constant, with a value of the local speed of sound. Comparison between calculated and experimental data for atomization of some different aluminium alloys in close-coupled atomizers and to Lubanska's correlation for mean drop sizes, yields heavily scattered results: a reasonable explanation for this behaviour has not been given.

A two-stage approach to describe primary and secondary fragmentation processes in a close-coupled atomizer has been derived by Liu (1997). First, a single phase, transonic and turbulent gas flow field is simulated in the atomizer. As a result, the gas velocity distribution and the location of recirculation areas are introduced as boundary conditions for the fragmentation model. The Reynolds-averaged Navier–Stokes (RANS) equations, Eq. (3.1), are solved using a conventional k–ε turbulence model. The grid system uses non-equidistant resolutions of the grid cells, especially in the area close to gas outlets. Figure 4.42 illustrates the resulting streamlines (left side) and the velocity vector distribution of the gas flow (right side). The effect of feedback of the moving fluid on gas flow behaviour is not taken into account (neglecting coupling effects between phases). Based on experimental observations of fragmentation processes in close-coupled atomizers (Ünal, 1987, 1988), it is assumed that the emerging fluid exits the tundish as a radial film, at the lower side of the nozzle tip, and is directed towards the gas jet delivery area. This effect is due to the so-called aspiration pressure of this type of atomizer. The surface velocity of the liquid film v_f and the film thickness τ_f in this area are derived from a Couette flow analogy:

$$\tau_f = \left[\frac{\mu_l}{\mu_g} \frac{\dot{m}_l}{\pi \rho_l r} \left(\frac{dv_g}{dz} \right)^{-1} \right]^{0.5}, \tag{4.51}$$

$$v_f = \left[\frac{\mu_g}{\mu_l} \frac{\dot{m}_l}{\pi \rho_l r} \left(\frac{dv_g}{dz} \right) \right]^{0.5}. \tag{4.52}$$

At the outer edge of the wetted nozzle tip, the liquid film is attacked by the gas flow field and is reflected into the main gas flow direction. The liquid, from this point on, moves as a free concentric ring in the direction of the gas streamlines. After a certain distance, the liquid film fragments into droplets. The characteristic distance from the atomizer to the fragmentation point is derived from empirical data:

$$L_f = 1.23 \, \tau_f^{0.5} \text{We}_f^{-0.5} \text{Re}_f^{0.6}. \tag{4.53}$$

The Weber number We_f and Reynolds number Re_f of the film are derived using the local film thickness τ_f and relative film velocity u_f as: $\text{We} = \tau_f \rho_g u_{\text{rel}}/2\sigma_l$; $\text{Re} = \tau_f \rho_l u_f/\mu_l$. As a result of the basic fragmentation process, droplets are produced at a characteristic drop size (here: Sauter mean diameter, SMD $d_{3.2}$) based on fitting to a semi-empirical model:

$$d_{3.2} = \frac{12\sigma_l}{\rho_l u_{\text{rel}}^2 \Big/ \left[1 + 1 \Big/ \left(\varepsilon \frac{\dot{m}_g}{\dot{m}_l} \right) \right] + 4 \frac{\sigma_l}{\tau_f}} \tag{4.54}$$

and

$$\varepsilon = \frac{1.62}{u_g^{1.3} \left(\frac{\dot{m}_g}{\dot{m}_l} \right)^{0.63} \mu_l^{0.3}}. \tag{4.55}$$

The droplets from this primary fragmentation process may be further disintegrated into smaller droplets depending on the local relative flow conditions. For this secondary disintegration model, the results of investigations from Hsiang and Faeth (1992; see Section 4.3.3) are used.

The empirical input of this model within close-coupled atomizers (Liu, 1997) is relatively high. However, comparison of Liu's simulated results with experimental data gives reasonable agreement for the mean droplet size. This comparison is illustrated in Figure 4.43. The cumulative drop-size distribution shows that at the upper end of the spectrum, that a significant amount of big droplets is missing in the calculation. This may be due to the neglected two-way coupling effect between the movement of the droplets and that of the gas phase. Here, due to neglect of the gas momentum sink for acceleration of droplets, the gas flow velocity in the fragmentation area has possibly been overestimated.

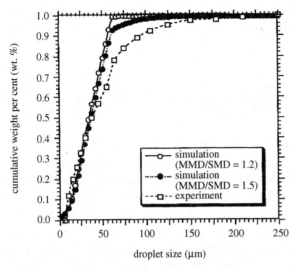

Fig. 4.43 Comparison of calculated and measured particle-size distributions (Liu, 1997) SMD, Sauter mean diameter; MMD, mass median diameter

Another model for fragmentation of a metal melt in a close-coupled atomizer has been introduced by Ting *et al.* (2000). This evaluates the effect of aspiration pressure on the flow of gas and liquid. In this way, flow in the atomization gas results in varying pressure distribution at the melt outlet. Based on dynamic assumptions for the gas flow field near the nozzle, for increasing gas pressures, Ting *et al.* conclude that with increased gas pressures an oscillating gas flow field may occur. This effect will result in the temporal occurrence of central orthogonal shock structures near the nozzle. Based on this observation, a fragmentation model is derived which predicts, for gas pressures exceeding a configuration-dependent threshold value, a pulsating melt flow field at the nozzle tip: a pulsating atomization process is, therefore, predicted.

General fragmentation modelling approaches

Several contributions attempt to derive a more general description of drop-size distributions resulting from atomization processes. These are based, for example, on statistical approaches and on tools which derive a probability density function in relation to the process to be analysed. The maximum entropy formalism (MEF) is such a statistical tool that delivers partial information for a specific process (based on a number of compulsatory conditions) by transformation into a suitable distribution function. It is a statistical tool and does not contribute any physical aspects to the analysis of fragmentation processes. Fundamental developments in the application of MEF to the analysis of liquid fragmentation processes and derivation of the resulting droplet-size distributions in sprays have been done, for example, by Sellens and Brzustowski (1985) and Li and Tankin (1987). Their approaches have been further developed by Ahmadi and Sellens (1993) and Cousin and Dumouchel (1997). The fundamental basis of these works is the description of a probability density function of the particle-size distribution $q_r(d)$ from the Shannon entropy (uncertainty) (see Shannon and Weaver, 1949):

$$\text{Sha} = -k_S \int_0^\infty q_r(d) \ln(q_r(d)) \, dd, \tag{4.56}$$

where k_S is a constant, to be determined by the dimensionality of the specific process under investigation.

The MEF allows anticipation of the probability density function depending on a number of mathematical constraints. These constraints are based on a number of known conservation properties of the process and their distribution, and are equivalent to moments of different orders of the distribution. Without further limitations, a number of different probability density functions will fulfil a given set of constraints. The MEF describes the best-fit solution as a function of the point at which entropy reaches a maximum value.

For description of the particle-size distribution in a spray in terms of the density distribution, the set of constraints can be described by the normalization criteria and the conservation condition (Cousin and Dumouchel, 1997):

$$\int_{d_{\min}}^{d_{\max}} q_0(d) \, dd = 1, \tag{4.57}$$

$$\int_{d_{\min}}^{d_{\max}} q_0(d) \, d^p \, dd = d_{p0}^p, \tag{4.58}$$

where $d_{\min}$ and $d_{\max}$ represent the minimum and maximum droplet diameter in the spray and p and d_{p0} (characteristic droplet diameter of the distribution, pth moment of the droplet-size distribution) are predefinable input parameters. Without further information on the process to be characterized, a constant density number of the droplet-size distribution for all particle-size classes is achieved. Maximization of the Shannon entropy based on the above constraints leads to the required particle-size distribution in the spray:

$$q_0(d) = e^{(-a_0 - a_1 d^p)}, \tag{4.59}$$

containing the Lagrangian multiplication factors a_0 and a_1. Cousin et al. (1996) describe the resulting density number based on the particle-size distribution of a pressure-swirl atomizer in this formalism as:

$$q_0(d) = \frac{p^{(p-1)/p}}{\Gamma_S\left(\dfrac{1}{p}\right) d_{p0}} e^{\left(\frac{-d^p}{p d_{p0}^p}\right)}, \tag{4.60}$$

containing the Gamma function Γ_S. The determination of input parameters like p and d_{p0} is either based on specific singular measurements (Ahmadi and Sellens, 1993) of moments of a measured spray particle-size distribution or by linear stability analysis of the fragmentation process (Cousin and Dumouchel, 1997). For an atomization process where the principal disintegration mode is constant (i.e. a constant fragmentation mode), the mean or characteristic particle size and the particle-size distribution width can be determined.

Further developments of MEF for spray and atomization processes are to be found in Malot and Dumouchel (1999) or Prud'homme and Ordonneau (1999), and for application in ultrasonic liquid atomization see Dobre and Bolle (1998).

Application of MEF to the atomization of liquid metals or spray forming applications has not been performed so far. Despite the empirical input required, use of this approach is promising.

Direct numerical analysis and simulation of phase boundary dynamics

Based on the discussion in Section 4.1.2 of the contour of the emerging liquid jet from an atomizer nozzle, direct numerical simulation of the dynamics of cylindrical fluid jets and phase boundaries may be performed. For physical modelling of this process, two important aspects have to be taken into account:

- the surface tension at the liquid/gas phase boundary has to be regarded as a boundary condition and
- localization of the transient phase boundary and its movement.

In addition to these two constraints, the excitation and deformation of a liquid jet in a coaxial gas flow needs to be discussed first. The calculations for this model are based on a two-dimensional approach. Therefore, only symmetrical perturbations of the liquid surface are analysed. Higher instability modes of the movement of the liquid surface are not analysed in this way. Therefore, the calculation is simplified and does not reflect all the deformation features encountered in reality in liquid jet. The surface tension model of Brackbill *et al.* (1992) is used in these calculations.

Figure 4.44 simulates the actual fluid jet contour at an external excitation wavelength of $(2d_i)$ and for the case of a coaxial gas velocity of 100 m/s. Movement and initiation of surface perturbations on the jet interface can be described in this way. Wave growth must be initiated by external excitation of the fluid jet at the entrance into the calculation area. Without external excitation, numerical instabilities overlay the solution. The reason for this is that the grid system used for this application is too coarse.

Recent investigations (Klein *et al.*, 2002; Lafaurie *et al.*, 1994; Lozano *et al.*, 1994; Mayer, 1993; Zaleski *et al.*, 1995) describe the complex phase boundary behaviour between gas and fluid, including the primary fragmentation process in atomization, based on a direct multiphase flow simulation. Rapid developments in the area of computer hardware, as well as in physical and numerical development of multiphase fluid models, allow completely three-dimensional numerical analyses, from initial liquid jet deformation to primary fragmentation of the liquid jet within the atomization process (Zaleski and Li, 1997). Based on the methods introduced in Section 4.1.2, such as the volume of fluid (VOF) method (Nichols *et al.*, 1980), in combination with surface reconstruction methods like the piecewise linear interface calculation (PLIC; Li, 1996), sufficient grid resolution of these calculations can be realized. Fragmentation and the simultaneous coalescence of fluid elements are described by these

fluid jet contour

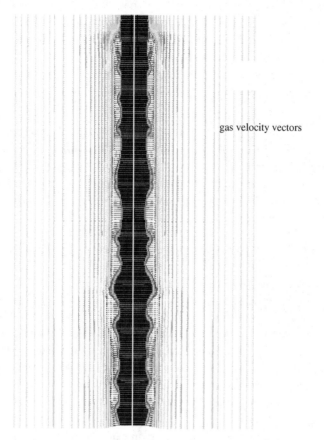

gas velocity vectors

Fig. 4.44 Calculation of liquid jet wave motion (symmetry mode)

models. Three-dimensional calculation is mainly used to observe excited fragmentation of a planar liquid sheet in a parallel gas flow field: as will be described next.

Figure 4.45 shows the fragmentation sequence from Zaleski and Li's (1997) simulation. The calculation has been performed for a liquid Reynolds number Re = 200, and a liquid Weber number of We = 300, at an artificially low-density ratio between the liquid and gas of $\rho_l/\rho_g = 10$ (at higher density ratios no stable solution can be achieved). From the tip of the evolving wave, a finger type ligament is formed, which is stripped off in a later stage from the main liquid jet and is deformed and stretched into the main flow direction to form a cylindrical fluid element. This unstable fluid element is fragmented later as a result of capillary instabilities, resulting in smaller fragments of different sizes. From this simulation, a primary fluid ligament results that is orientated in the direction of the main flow. In contrast to this result, the classical aerodynamic fragmentation model of Dombrowski and Johns (1963; see Figure 4.38) obtains a fluid ligament that is orientated perpendicular to the main flow direction.

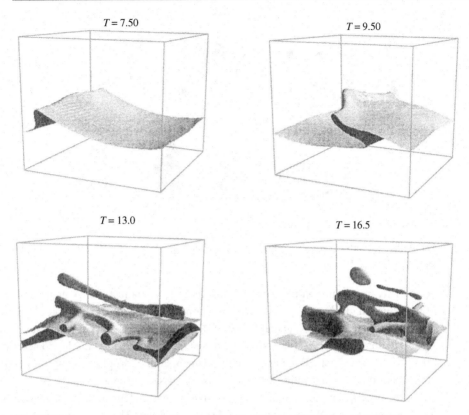

Fig. 4.45 Fragmentation simulation of a plane liquid sheet (Zaleski and Li, 1997)

Direct numerical calculation (only in two dimensions) of the dynamics of phase boundaries of a planar liquid sheet is performed by Lafaurie *et al.* (1998) and Klein *et al.* (2002). They discuss some of the main effects of liquid parameters, like the density ratio, the viscosity ratio and capillary effects, on the primary fragmentation mechanism, based on simulation results and comparisons with experimental results.

Despite the limited capabilities of direct numerical simulations in terms of the possible spectra of process parameters (stable solutions are to be achieved only with simplified parameters for a small range of liquid conditions), discussion of such simulation results will give a much deeper insight into the physics of fragmentation processes in the near future. This approach enables derivation of some important fragmentation process parameters, and the possibility of calculating a great number of parameter sets will help to optimize atomization processes (for example, advanced atomizer configurations or atomization assisted by external or internal excitation of the liquid jet).

4.3.3   Secondary atomization

Following the primary fragmentation process of a liquid jet into droplets and/or fluid ligaments, further break-up of these liquid elements depending on the local flow structure may

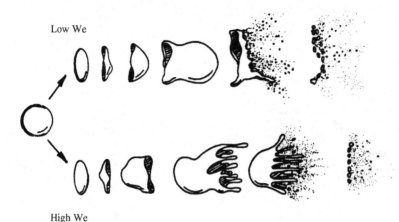

Low We

High We

Fig. 4.46 Secondary droplet atomizations: bag- and stripping-type break-up of a droplet (Bayvel and Orzechowski, 1993)

occur. This fragmentation process is called secondary atomization. A general discussion of the principal mechanisms and results of secondary liquid atomization processes is to be found in Faeth (1990), Hsiang and Faeth (1992, 1993), Bayvel and Orzechowski (1993), as well as Sadhal *et al.* (1997). In this context, secondary fragmentation due to aerodynamic effects is described here; other effects like fragmentation of droplets during collision or the collision of droplets and solid surfaces will be discussed in separate chapters.

For the description of secondary atomization of liquid elements, fragmentation models are used which are different from those discussed earlier. These models are divided into two main modes. Both are initiated by perpendicular flattening of the element due to the asymmetric pressure distribution acting on the surface of the liquid in the gas flow field at a relative velocity between the gas and the droplet. Another principal mode of disintegration is valid for asymmetric flow fields, as in shear flows, and will not be referred to here. First, one needs to distinguish between two basic modes of fragmentation: bag-type break-up and stripping-type break-up. In the former, the flattened fluid element forms an inner membrane which is drawn out from inside the liquid element. In a subsequent stage, this membrane breaks up into a series of relatively small droplets. This model is illustrated in Figure 4.46 (Bayvel and Orzechowski, 1993), which is valid for:

$$We = \frac{\rho_g u_{rel}^2 d_p}{\sigma} > 12. \tag{4.61}$$

In the case of stripping-type break-up, droplets are produced by stripping from the edge of the fluid element. This behaviour is observed for

$$\frac{We}{\sqrt{Re}} > 1, \tag{4.62}$$

and is also illustrated in principle in Figure 4.46.

A third mode of secondary atomization may be introduced by elongation of the fluid element in an external flow field with velocity gradients. Here, the elongated particle fragments

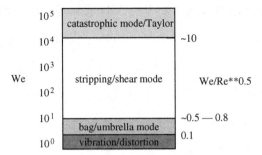

Fig. 4.47 Modes of secondary break-up of droplets, Reynolds number Re, aerodynamic Weber number We (Delplanque and Sirignano, 1994)

into small droplets due to Rayleigh instabilities in the liquid fibre. A number of subdivisions to these main modes of secondary atomization have been described (Bayvel and Orzechowski, 1993; Sadhal *et al.*, 1997). No general models valid for the entire range of processes and liquid parameters are available: additional research is needed in this area.

The dependence of the main modes of secondary droplet break-up Weber and Reynolds numbers, is illustrated in Figure 4.47.

Secondary drop fragmentation models in spray analysis have been successfully used by O'Rourke and Amsden (1987; the Taylor analogy break-up (TAB) model), which is frequently cited, and by Reitz and Diwarkar (1987) for analysis and simulation of Diesel injection spray drop break-up.

An important criterion for droplet break-up is (besides the critical Weber number limit, as a steady limit) the dynamic behaviour of the liquid. Therefore, minimum interaction and deformation times, as well as a minimum fragmentation time, need to be determined. Without regard to such dynamic effects, the real fragmentation rate will be overestimated in static models. The fragmentation delay has been formulated by Nigmatulin (1990) and has been used, for example, in the work of Berthomieu *et al.* (1998) as:

$$\frac{t_{\text{fragmentation}}}{t^*} = 6(1 + 1.2\text{La}^{-0.37})(\log \text{We})^{-0.25}. \tag{4.63}$$

The dynamic scale of fragmentation is derived in terms of a dimensionless time t^* and the Laplace number La:

$$t^* = \frac{d_p}{u_{\text{rel}}}\sqrt{\frac{\rho_l}{\rho_g}}, \quad \text{La} = \frac{\rho_l \sigma d_p}{\mu_l^2}. \tag{4.64}$$

Secondary processes during the atomization of molten metals are important, though in this case the droplet Weber number is usually very low because of the high surface tension of molten metals.

Secondary break-up in spray forming applications

A secondary melt atomization break-up model for spray forming applications has been derived and applied by Markus and Fritsching (2003). From an initial classification of droplet break-up mechanisms, several investigations describe the resulting droplet diameters. The

TAB model of O'Rourke and Amsden (1987) derives the resulting Sauter mean diameter (SMD) $d_{3.2}$ for very low,

$$d_{3.2} = \frac{3}{7} d_p \quad \text{for} \quad \text{We} \to \text{We}_{\text{crit}}, \tag{4.65}$$

and high Weber numbers:

$$d_{3.2} = 12 \frac{d_p}{\rho_g u_{\text{rel}}^2} \quad \text{for} \quad \text{We} \to \infty. \tag{4.66}$$

The initial droplet diameter is d_p.

Hsiang and Faeth (1992) derived an empirical correlation between the SMD and the Ohnesorge (Oh) and Weber numbers for conventional liquids at various viscosities

$$\frac{d_{3.2}}{d_p} = 1.5 \text{Oh}^{0.2} \text{We}_g^{-0.25} \tag{4.67}$$

in the range $\text{Oh}^{0.2} \text{We}_g^{-0.25} = 0.05 \ldots 0.4$. In the case of shear break-up, the resulting droplets, with diameter d, have been found to follow a normal distribution path:

$$f(d) = \frac{\sqrt{d/d_{50}}}{2\sqrt{2\pi}\sigma d} e^{-\frac{1}{2}\left(\frac{\sqrt{d/d_{50}} - \mu}{\sigma}\right)^2}, \quad \mu = 1, \quad \sigma = 0.238. \tag{4.68}$$

The normal particle-size distribution has been taken by Schmehl (2000) to indicate bag- and multimode-type break-up mechanisms.

The dynamics of secondary droplet break-up depend on the aerodynamic Weber number. At critical conditions, at first the droplet deforms to a disc and then breaks up into smaller droplets, as illustrated in Figure 4.48. The characteristic break-up time is t^* and the deformation time is $t_{\text{def}} = 1.6t^*$ (Hsiang and Faeth, 1992). The data of Samenfink et al. (1994) may be used for the total break up time scale t_b as:

$$t_b = kt^*, \tag{4.69}$$

where k is an empirical constant. Compared with the break-up time, the spheroidization time t_s of a droplet after break-up is small and may be neglected in a model (Nichiporenko and Naida, 1968):

$$t_s = 0.88 \frac{\mu_l d_p}{\sigma}. \tag{4.70}$$

As an example, the spheroidization time of a molten steel droplet of 100 μm size is $t_s = 250$ ns and therefore $t_s \ll t_b$.

The computational approach described here for the secondary break-up process in molten metal atomization and spray forming is based on a two-dimensional representation of the fragmentation zone. The description of the secondary break-up process for complex nozzle geometries accounts for the total bulk of all molten particles. *Every* particle in the break-up zone of the spray is tracked and calculated. Therefore, within the overall model of the break-up process, the following have been taken into account:

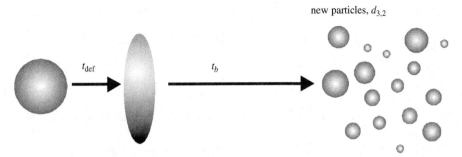

new particles, $d_{3,2}$

Fig. 4.48   Sequential deformation and break-up of a droplet

- the probability of drop break-up;
- the outcome of drop fragmentation;
- a model for droplet acceleration and simultaneous solidification/cooling of all particles; and, eventually,
- a model for possible drop collisions (coalescence or fragmentation).

For each particle, instantaneous thermal and kinetic situations are analysed, and a decision is made as to whether the break-up procedure is initialized or not. The break-up model terminates when:

- either the critical Weber number is not achieved,
- or the droplet has been cooled down to the point at which nucleation (including an appropriate undercooling model, see Section 5.1.2) first occurs (see solidification modelling in Section 5.1.2). Here it is assumed that the viscosity of the droplet rapidly increases and prevents the droplet from breaking up further.

The computational approach is one-way coupled (i.e. the particles have no effect on the behaviour of the gas phase) and drop collisional effects have been ignored. The prescribed gas field variables are: the mean gas velocity in axial and radial directions u_g, v_g; the turbulent kinetic energy k of the gas, and from there the turbulent velocity components u'_g, v'_g, and the gas temperature T_g. The distribution of gas parameters near the atomizer nozzle is taken from experimental investigations or gas flow simulations (see Section 4.2). The derivation of models for description of droplet kinetics and thermal behaviour along a droplet trajectory (plus a suitable solidification model) will be described in Chapter 5.

Results for the atomization of a nickel-based superalloy are shown here for secondary break-up in a standard free-fall atomization nozzle arrangement for an atomization pressure of $p_g = 0.4$ MPa, a gas, metal mass flow ratio (GMR) of $= 0.68$, and a molten metal jet diameter of $d_0 = 4$ mm. Starting with the analysis of linear metal jet stability prior to break up (see Section 4.1.2), the initial ligament/particle diameter and the initial variance of the distribution have been estimated as the first input parameters for the secondary break-up model.

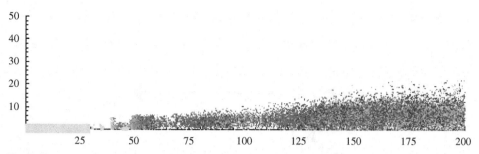

Fig. 4.49 Result of secondary break-up simulation accounting for all particles (shading is used to represent the thermal state of the droplets; Markus and Fritsching, 2003)

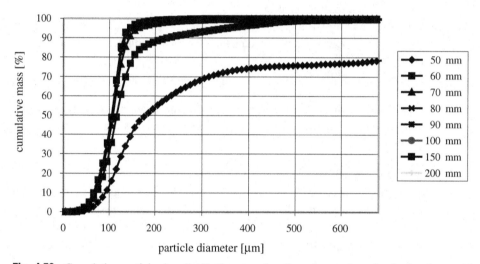

Fig. 4.50 Cumulative particle-size distribution as a function of secondary droplet break-up with respect to distance from the primary break-up point (Markus and Fritsching, 2003)

Figure 4.49 shows the simulation results of secondary droplet break-up in the atomizer at distances between 0 and 200 mm from the nozzle: the instantaneous spatial distribution of droplets in a two-dimensional slice through the spray is shown. Shading of the droplets is used to depict individual particle temperatures. The continuous intake length of the metal jet prior to break-up is assumed to be $L = 30$ mm. Most particles are produced in a secondary fragmentation zone, a short distance ($\Delta x \sim 50$ mm) from the region of primary break-up. Once break-up is terminated (due to undercritical kinetic conditions or achievement of droplet solidification), the droplets are only further accelerated slightly, before being spread out due to turbulence.

The resulting cumulative particle-size distribution does not in the initial spray shifts towards smaller particles with increasing distance from the nozzle. At the start of the disintegration process, especially, only a few large particles contain most of the melt mass (see Figure 4.50). The secondary break-up process is almost complete at a nozzle distance of 80 mm, as the droplet-size distribution does not change further at increasing nozzle distance. Therefore, in this case, the break-up region is located within 50 mm of the spray.

The results of the break-up simulation at this point may be used as input parameters for two-phase simulation of spray behaviour (e.g. including coupling effects and other submodels, see Chapter 5). From the break-up model, the initial droplet-size distribution, the spray angle and the mass flux distribution after fragmentation, may be used to derive the spray model following.

5   Spray

Analysis of turbulent multiphase flow in a spray is of major concern during numerical modelling and simulation, as the turbulence is responsible for a number of subprocesses that affect spray forming applications. These result from coupled transport between drop and gaseous phases, and from extensive transfer of momentum, heat and mass between phases due to the huge exchange area of the combined droplet surface. Physical modelling and description of these exchange and transport processes is key to the understanding of spray processes.

In spray forming, especially, the thermal and kinetic states of melt particles at the point of impingement onto the substrate, or the already deposited melt layer, are of importance. This is the main boundary condition for analysis of growth, solidification and cooling processes in spray formed deposits. These process conditions finally determine the product quality of spray deposited preforms. By impinging and partly compacting particles from the spray, a source for heat (enthalpy), momentum and mass for the growing deposit is generated. The main parameters influencing successful spray simulation in this context are:

- the local temperature distribution and local distribution ratio between the particles and the surface of the deposit,
- particle velocities at the point of impingement, and
- the mass and enthalpy fluxes (integrated rates per unit area and time) of the compacting particles.

Distribution of these properties at the point of impingement is determined mainly by the fragmentation process and by the transport and exchange mechanisms in the spray.

An important point for further process developments within spray forming is the application of suitable measurement and in-process control equipment for *in-situ* detection of spray particle properties. However, due to the harsh environment within metal spray applications and the extreme process parameters to be covered, the determination of particle thermal properties in spray forming relies mainly on numerical data. Modelling approaches can only derive the thermal state of the particles in terms of temperature, plus the state of solidification of the particles. Here multiphase flow simulation at different degrees of complexity may be applied. The term complexity in this context stands for:

- increasing effort needed for simulation and numerical calculations,
- the possibility of simultaneously analysing a higher degree of interactions between physical processes and providing significantly greater description of subprocesses.

Therefore, the term complexity is directly coupled to the demand for process parameter details and the properties to be derived.

Different stages of complexity of spray analysis can be reflected in the number of physical dimensions (in space and time) taken into account in the modelling approach and, in addition, by the number of coupling parameters (see, for example, Crowe (1980)):

(1) One-dimensional spray calculations analyse the behaviour of particles at the centre-line of the spray only. The behaviour of these representative particles at the spray centre-line is defined to be characteristic for the spray as a whole. Coupling between phases in this modelling approach is not regarded.
(2) One-dimensional and quasi-two-dimensional approaches extrapolate the behaviour from the centre-line of the spray towards the spray in its entirety, based on suitable models or known empirical correlations.
(3) Full two-dimensional or three-dimensional models analyse and describe the multi-dimensional multicoupled spray behaviour directly.

Within one-dimensional and quasi-two-dimensional models for spray analysis, the effect of the gas phase (i.e. the gas velocity distribution – and in the case of metal atomization the gas temperature distribution) on the centre-line of the spray is prescribed, and is based on:

- measurement results and experimental investigations of the multiphase flow field in the spray,
- simplified analytical solutions or measurements of the single-phase flow (without atomization),
- numerical simulation results derived for comparable processes and transferred to the specific application.

Adopting this approach, the behaviour of individual droplets in the spray can be determined by tracking droplets of different size and properties through the spray. The effect of particle behaviour on the state of the continuous gas phase is not regarded: this analysis feature is only one-way coupled.

A quasi-two-dimensional model at first analyses the behaviour of the particle on the centre-line of the spray (or on another characteristic particle trajectory) without coupling. By extrapolation (weighted or unweighted) of the results from the centre-line, based on a given local distribution of, for example, drop sizes and/or drop-mass fluxes, the radial distribution of individual thermal and kinetic particles is derived, which is finally integrated and correlated with the state of the spray at impingement (Pedersen, 2003; Zhang, 1994).

Full two-dimensional or three-dimensional coupling determines the interaction and exchange of, for example, momentum, mass and thermal energy in the conservation balance for both liquid and gaseous phases. For example, local acceleration of particles in a gas flow field results in a momentum sink for the gas phase, which is described by sink terms in the relevant momentum conservation equations of the gas phase (Grant *et al.*, 1993a, b).

The multicoupled simulation of turbulent dispersed multiphase flow, containing a continuous phase (i.e. gas) and a dispersed phase (i.e. droplets or particles), is based on two different modelling concepts:

- the Eulerian/Lagrangian approach, and
- the Eulerian/Eulerian approach.

In common, both concepts use conservation equations for mass, momentum and energy, plus a suitable turbulence model within a Eulerian reference frame, to analyse the behaviour of the gaseous phase. The types of models mainly used in this context describe the stress tensor of additional turbulent stresses of the continuous phase, based on a Boussinesq approximation with a turbulent viscosity v_t. This turbulent viscosity is described by the local flow field configuration, and is based on conservation equations for the turbulent kinetic energy k and its dissipation rate ε.

More recent models for turbulent transport behaviour of the continuous phase dissolve the wave structure anisotropically into different scale wavelengths, for example in:

- direct numerical simulation, DNS (see, for example, Albrecht *et al.*, 1999);
- large eddy simulation, LES (see, for example, Edwards, 1998; Bergström *et al.*, 1999);
- approaches based on probability density functions (pdf) of gaseous properties (see, for example, Rumberg and Rogg, 1999);
- discrete vortex methods (see, for example, Crowe *et al.*, 1996).

These modelling approaches for continuous phases within a dispersed multiphase flow simulation provide extremely useful basic data on physical transport mechanisms and properties, but have been limited up to now to simple base geometries and boundary conditions.

The main differences between dispersed multiphase flow simulations are due to proper description of the particulate phase. Here, either a particle tracking approach for individual particles or for particle clusters is used (Lagrangian approach) or the dispersed phase is treated as a quasi-second fluid with spatially averaged properties (Eulerian approach). The advantages and disadvantages of both modelling approaches for dispersed multiphase flow problems are discussed in a number of conferences and publications (see, for example, Crowe *et al.*, 1996; Durst *et al.*, 1984; Sommerfeld, 1996).

Eulerian/Lagrangian approach

The Eulerian/Lagrangian approach is closely related to the direct intuitive approach which one would make to analyse the behaviour of dispersed multiphase flow. Individual particles are tracked through the flow field and their local interaction with their specific surrounding is analysed on the scale of the droplet size (point source assumption). In this context, a number of significant physical effects may be incorporated, such as microscopic effects of the flow around individual particles, interaction between neighbouring particles and the interaction of particles with surfaces. These effects may be important under specific circumstances for some types of multiphase flows. Some disadvantages of this approach may result from the necessary averaging of individual particle properties to derive the intermediate spray structure at a specific position. Here a great number of particles need to be tracked in order to obtain a statistically meaningful average. Coupling is achieved by including the local sources/sinks from the particles as point sources in the continuous equations. The main method for coupled analysis of dispersed multiphase flow is the particle-source-in-cell

(PSI) algorithm of Crowe *et al.* (1977). Several models within spray forming applications based on the Eulerian/Lagrangian approach have been published (see, for example, Grant *et al.*, 1993a, b; Bergmann *et al.*, 1995; Fritsching, 1995; Fritsching and Bauckhage, 1994a).

Eulerian/Eulerian approach

In the Eulerian approach, the dispersed phase is treated as a quasi-second fluid with spatially averaged properties. Integration of microscopic effects on the droplet-size scale, for example, for interaction of droplets with each other, is much more complicated within Eulerian/Eulerian models than within the Lagrangian/Eulerian approach. Here, for example, analysis of dispersed phase turbulent properties and structures, as well as interactions based on stress tensor derivations of the dispersed phase, develop modelling problems. Such Eulerian/Eulerian modelling for two-phase jets has been discussed by Elghobashi *et al.* (1984). One advantage of this approach is the possibility of using almost identical numerical solvers and algorithms, as the basic equations for continuous and dispersed phases have identical structures. The drastically increasing effort of Eulerian particle models to describe dispersed multiphase flows containing polydispersed particle-size distributions (as in this case each individual drop-size class needs to be treated with a separate set of conservation equations) limits the wider use of this approach in technical simulation applications.

Derivation of the spray structure within a spray forming process based on an Eulerian/Eulerian simulation approach has been done by Liu (1990) and by Fritsching *et al.* (1991). In this model the turbulent properties of the dispersed phase have been directly coupled algebraically to the turbulence properties of the continuous phase, based on the turbulent viscosity concept. The particle-size distribution in the spray has been neglected, the particle sizes have been represented by the mean particle size of the spray only. Spray generation and spray development within molten metal atomization for a free-fall atomizer configuration has also been documented by Dielewicz *et al.* (1999). Here, the coupling of a transonic gas flow field and a two-dimensional spray is discussed.

5.1   Particle movement and cooling

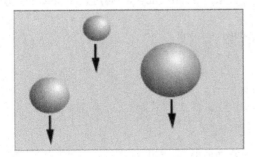

In this section some models, and their results, for particle movement and cooling along the in-flight particle trajectory during spray forming will be introduced. A fundamental

Table 5.1 *Force balance for the trajectory equation of a particle*

(1) inertial force	$F_t = -\rho_p V \dfrac{\mathrm{d}v_p}{\mathrm{d}t}$
(2) field force	
• gravity force	$F_g = \rho_p V g$
(3) pressure forces	
• buoyancy force	$F_a = -\rho_g V g$
• pressure gradients	$F_p = -\rho_g \dfrac{\mathrm{d}v_g}{\mathrm{d}t} V$
(4) fluid mechanics forces	
• resistance force	$F_w = -\dfrac{1}{2}\rho_g A_p c_w(\mathrm{Re})|v_p - v_g|(v_p - v_g)$
• added mass	$F_m = -\dfrac{1}{2}\rho_g V \left(\dfrac{\mathrm{d}v_p}{\mathrm{d}t} - \dfrac{\mathrm{d}v_g}{\mathrm{d}t}\right)$
(5) other forces	
• Basset history integral	$F_b = \dfrac{3}{2}d^2 \sqrt{\dfrac{\rho_g \mu}{\pi}} \left(\int_{-\infty}^{t} \dfrac{\frac{\mathrm{d}u_p}{\mathrm{d}t} - \frac{\mathrm{d}u_g}{\mathrm{d}t}}{(t-\tau)^{1/2}}\mathrm{d}\tau + \dfrac{(u_p - u_g)_0}{\sqrt{t}}\right)$

description of the behaviour of droplets or solid particles in fluids, the flow around particles and their analysis is given in Clift *et al.* (1978), Crowe *et al.* (1998), Sadhal *et al.* (1997) and Sirignano (1999).

5.1.1 Momentum transfer and particle trajectory equation

Newton's law presents the (at first one-dimensional) balance force for an individual spherical particle. The most relevant forces are listed in Table 5.1 based on the coordinate system in Figure 5.1. From this expression the position and displacement of a particle is:

$$u_p = \frac{\mathrm{d}x}{\mathrm{d}t}. \tag{5.1}$$

Most contributions for analysis of dispersed phase behaviour in a continuous flow field are based on the components listed by Maxey and Riley (1983) in the so-called Basset–Boussinesq–Oseen (BBO) equation.

The particle trajectory equation is derived from:

$$\Sigma F = 0. \tag{5.2}$$

This equation takes into account:

- particle inertia (1),
- the stationary force contributions from gravity (2) and
- buoyancy (3a), as well as
- the resistance force of the particle (4a).

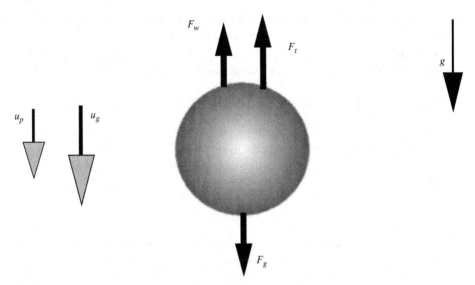

Fig. 5.1   Coordinate system for force-balancing for an individual particle

The other contributions listed describe:

- the force from the asymmetric pressure distribution around the particle due to the accelerated outer fluid (3b),
- the transient forces of added mass (4b), and
- the Basset history integral (5).

The added-mass term describes the contribution of the surrounding fluid, which needs to be accelerated together with the particle in the boundary layer of the particle. The Basset history integral takes into account the delayed development and establishment of the particle boundary layer in a transient flow field. Determination of this contribution takes into account the whole history of the particle trajectory from the starting point to its actual final position. This contribution is important in strongly transient or turbulent flow fields. The last term in the Basset history integral has been introduced by Reeks and McKee (1984) for the case of a finite particle starting velocity.

Dependent on the application and on the boundary conditions, further particulate force contributions and influences need to be included in the particle force balance, such as:

- forces in electrical or rotational fields (e.g. for analysis of particle separation in electrofilters or cyclones),
- turbulence effects (turbulent particle dispersion behaviour),
- particle/particle interactions in clusters or swarms of particles at higher concentrations,
- compressibility effects in transonic and supersonic flows,
- particle rotations (in shear flow or resulting from inclined drop impacts) resulting in lift forces,
- lift forces normal to the flow direction in shear flow fields, and
- other influences.

In the analysis of fluidic drops or solid particles in a gaseous atmosphere, the density ratio between gas and particles is typically sufficiently small ($\rho_g/\rho_p < 10^{-3}$). For molten metal droplets and solid metal particles this density ratio is even smaller. In this special case, the particle trajectory equation can be simplified with only a small loss of accuracy to:

$$m_p \frac{d\mathbf{u}_p}{dt} = m_p \mathbf{g} + \frac{1}{2}\rho_g |\mathbf{u}_g - \mathbf{u}_p|(\mathbf{u}_g - \mathbf{u}_p)c_d A_p. \tag{5.3}$$

This remaining force balance equation only takes into account particle inertia, gravity and resistance. The resistance (drag) coefficient c_d in most cases is taken from the analysis of a single spherical solid particle, which can be described in the range of Reynolds numbers Re < 800 by:

$$c_d = \frac{24}{\text{Re}}(1 + 0.15\text{Re}^{0.687}), \quad \text{Re} < 800. \tag{5.4}$$

Liquid droplets deviate from solid particles during interaction with gases due to the free mobility of the drop surface, leading to droplet deformation as well as to movement of the drop surface (and the correlated inner circulation of the droplet). In addition, the drop may oscillate. This transient behaviour of the drop surface may be correlated with surface tension forces. In the area of Stokes flow (Re < 1), the resistance coefficient of a fluidic droplet with mobile phase boundary and inner circulation may be expressed (see, for example, Clift *et al.*, 1978) as:

$$c_d = \frac{24}{\text{Re}}\left(\frac{1 + \frac{2}{3}\overline{\mu}}{1 + \overline{\mu}}\right), \quad \text{Re} < 1. \tag{5.5}$$

In this correlation $\overline{\mu}$ is the viscosity ratio of the inner fluid in the droplet to the surrounding fluid (μ_d/μ_g). A correlation for approximation of the resistance coefficient of a droplet with respect to ideal inner circulation and without boundary layer separation at the surface of the droplet has been derived by Chao (1962) and Winnikow and Chao (1965). This expression is valid for the limiting case of an ideal stationary, rising or sinking droplet based on boundary layer theory and is:

$$c_d = \frac{16}{\text{Re}}\left(1 + \frac{0.814}{\text{Re}^{0.5}}\right)\left\{\frac{2 + 3\mu_{\text{Tr}}/\mu_l}{1 + [(\rho_{\text{Tr}}/\rho_l)(\mu_{\text{Tr}}/\mu_l)]^{0.5}}\right\}, \quad 80 < \text{Re} < 400. \tag{5.6}$$

An overview of a number of other correlation functions for resistance coefficients for liquid droplets is given in Brander and Brauer (1993).

A number of approaches are to be found in the literature which analyse the flow field around a droplet simultaneously with the mass transfer behaviour of particles and droplets: determination is by experiment as well as by direct numerical calculation. For example, Brander and Brauer (1993) solved the problem of a stationary flow around a spherical non-deformable fluidic particle, which is moving in an unbounded fluid of infinite dimensions. For the theoretical investigation, the calculation is based on the coupled velocity and pressure distribution within and around the droplet. These distributions are derived from a finite element analysis of the momentum, mass and concentration conservation equations. The conservation equations are formulated for each phase (inner and outer) separately and are coupled by suitable coupling and transfer functions at the droplet boundary and free surface (see Section 4.3.2). In that contribution, especially, the onset and development of flow separations and vortices within the droplet and at the outer side of the droplet are discussed and related to local details of the flow field.

In the case of molten metal droplets one has to account for high surface tension values normally assigned to molten metals. These will result in small values of the Weber number of melt droplets within the spray. Therefore, considerations of droplet oscillations and deformations are of minor importance in this case within the spray, despite the occurrence of an area close to the fragmentation of the melt stream, where deformed droplets may be found frequently.

Analysis of droplet behaviour on the centre-line of the spray (one-dimensional)

Based on a one-dimensional analysis of the particle trajectory equation with only one-sided coupling (see Point 1 in Table 5.1), investigation of particle behaviour in sprays (e.g. for identification of different atomizer nozzle characteristics) is possible. In Fritsching and Bauckhage's (1987) analysis of interaction of particles and the surrounding gas phase with distance in the spray from single (pressure) and twin-fluid (gas) atomizers, the behaviour of droplets within a model fluid spray (water) has been investigated. Droplet velocities and sizes at the centre-line of the spray have been measured by phase-Doppler-anemometry (PDA) for comparison with the one-dimensional model. The nozzle types under investigation (as there are pressure and twin-fluid atomizers) and their respective dimensions have been chosen so as to yield comparable droplet sizes and velocity spectra within the spray.

For solution of the droplet trajectory using a one-dimensional non-coupled model approach, assumption of a prescribed velocity distribution in the gas phase is necessary, and needs to be prescribed for the centre-line of the spray. This important boundary condition for model evaluation has been determined from PDA measurements of droplet velocities of the smallest detected droplets in the spray, based on the assumption that the local gas velocity equals the measured droplet velocity for the smallest detected droplet-size class. This is the ideal case for drop movement, i.e. when the slip velocity between the gas and the droplets vanishes for smaller droplets at low gas acceleration rates because of their low inertia. As another necessary boundary condition, the initial (starting) velocity value of the droplet needs to be prescribed. For the twin-fluid nozzle it is assumed that the starting velocity of the droplets is equal to the liquid exit velocity from the nozzle. Therefore, its value has been determined by measuring the mass flow rate and the nozzle exit diameter. For the pressure atomizer it is more difficult to define a specific starting velocity for the droplets as here more pronounced multistage ligament fragmentation process of the liquid occurs. In an approximation, a constant starting velocity for all droplet-size classes has been assumed. This value has been achieved by extrapolating the measured droplet velocities of the biggest detected drop-size classes, from the first measurement location in front of the nozzle, backwards to the common starting position of the droplets near the nozzle.

In Figure 5.2 comparison between experimental and measured results, averaged over a certain drop-size class for the pressure atomizer used, are illustrated. The dashed/dotted

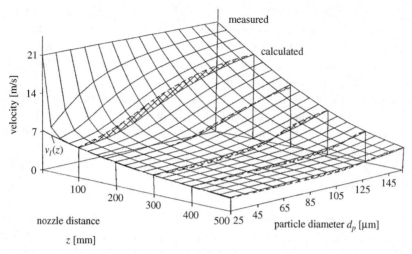

Fig. 5.2 Velocity distribution at the centre-line of a pressure atomizer (single fluid nozzle; Fritsching and Bauckhage, 1987)

line in the figure indicates the derived gas velocity distribution as a function of nozzle distance $v_l(z)$. The figure shows that the smaller droplets in the spray immediately lose their initial momentum once carried from the liquid delivery system. This initial momentum is immediately lost and transferred to the slow moving (entrained) gas. Further downstream, these smaller droplets move at almost the same velocity as the gas. The bigger droplets decelerate somewhat more slowly. Therefore, a continuous transfer of momentum from the droplets to the slower moving gas occurs. The velocity distribution exhibits strong gradients dependent on droplet size in the vicinity of the nozzle, while at increased nozzle distances the velocity distribution for all droplet sizes is equalized.

In the spray of the twin-fluid atomizer, a change in sign of the direction of momentum transfer occurs. This can be seen in Figure 5.3. Close to the atomizer, the faster moving atomizer gas accelerates all droplet sizes. This is the direction of momentum transfer from the gas to the slower particles. Therefore the gas looses a significant amount of kinetic energy within a small distance from the atomizer. At a certain distance from the atomizer, the direction of momentum transfer is reversed. At greater nozzle distances, all droplets move faster than the gas; therefore the direction of momentum transfer is now from the droplets to the gas. The droplets in this area accelerate the gas. The point where the momentum transfer changes direction depends on droplet size. While smaller droplets already exhibit this change of momentum transfer closer to the atomizer, due to their smaller inertia, bigger particles experience a change in their direction of momentum transfer at somewhat greater distances from the nozzle.

By comparing the characteristics of both nozzle types under investigation, from this simple one-dimensional model evaluation, one can already observe remarkable differences in the technical development of spray forming processes. The aim of the application is to achieve almost identical droplet velocities at a specific distance from the atomizer

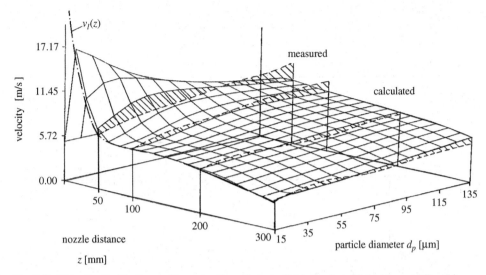

Fig. 5.3 Velocity distribution at the centre-line of a twin-fluid atomizer (Fritsching and Bauckhage, 1987)

(e.g. for coating applications): for the twin-fluid atomizer under investigation, this condition is achieved much nearer to the apparatus. In addition, reversal of the direction of momentum transfer has to be recognized as an important feature of twin-fluid atomizers.

In the field of spray forming simulations, quasi-two-dimensional model calculations have been performed, for example, by Zhang (1994). Here, the velocity behaviour of the droplets will be illustrated first. The gas velocity distribution has been assumed from measurements of gas velocities at the centre-line of the free gas jet flow (produced by an atomizer but without atomization) by Uhlenwinkel (1992). For the measurement, the original atomizer nozzle system of a spray forming device has been used. The area under investigation is located within $100 < z < 800$ mm of the nozzle. As discussed in Chapter 4, for a typical free-fall atomizer, disintegration of the liquid occurs approximately 100 mm below the atomizer. Therefore, the investigated area covers the entire spray region, i.e. from atomization to spray impingement onto the substrate. Based on comparable flow field configurations of a turbulent free single-phase jet, the velocity distribution in the atomizer gas flow field has been divided into two regions. Near the nozzle, a plug flow profile of gas velocities has been assumed, with constant velocity:

$$u_g = u_{g0} = \text{const}, \quad 100 < z < 300 \, \text{mm}. \tag{5.7}$$

At greater distances from the atomizer, the decrease in gas velocity at the centre-line has been assumed to occur in an exponential manner:

$$u_g = A_l(p_g^*)^{A_2} \left(\frac{z}{d_g} \right)^{-A_3}, \quad z > 300 \, \text{mm}. \tag{5.8}$$

Table 5.2　*Velocity correlation parameters*

d_g [mm]	A_1	A_2	A_3
20.4	0.0123	0.53	0.95

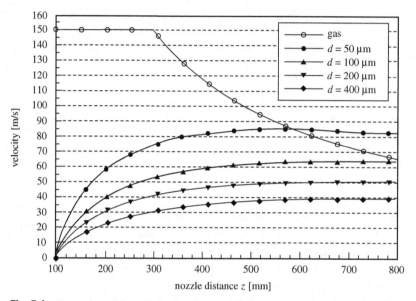

Fig. 5.4　Gas and particle velocity distribution at $v_{g0} = 150$ m/s (Zhang, 1994)

The gas velocity depends on the relative gas prepressure p_g^*, where $p_g^* = p_{g,abs} - p_u/p_u$, the characteristic nozzle dimension of the free-fall atomizer d_g and the correlation coefficients from Table 5.2.

The calculated particle velocities, at the centre-line of a spray of different-sized metal particles, during atomization of a steel melt is shown in Figures 5.4 and 5.5. The examples have been calculated for different gas pressure levels during atomization, therefore assuming different gas velocities near the atomizer. As previously mentioned, in this example the gas velocity is taken as constant near the nozzle. The particles start at a distance of 100 mm below the atomizer, all particles have almost zero starting velocity at the fragmentation point, respectively their starting location. Dependent on droplet inertia, the particles of different size show different acceleration behaviour in the gas. The typical reversal of momentum transfer direction within gas atomization, mentioned earlier, can only be observed for particles with sizes $d_p < 100$ µm. For bigger particles the flight distance is too small for them to experience a change in their momentum transfer direction. These particles always exhibit smaller velocities than the gas velocity until the point of droplet impingement onto the substrate.

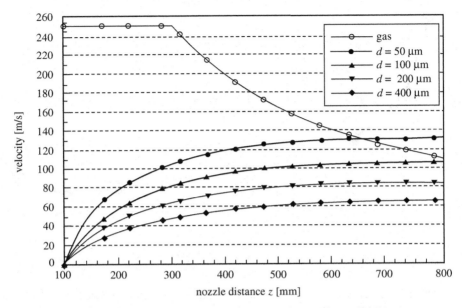

Fig. 5.5   Gas and particle velocity distribution at $v_{g0} = 250$ m/s (Zhang, 1994)

Turbulent dispersion of particles in the spray

Due to the turbulent structure of the gas flow field, the particles may deviate from the deterministic trajectory, which is governed by the mean gas flow properties. This effect is called turbulent particle dispersion. The way in which this problem may be tackled has been outlined, for example, in Sommerfeld (1996). The model is based on the estimation of instantaneous gas velocity **u** from Reynolds' decomposition:

$$\mathbf{u}_g = \bar{\mathbf{u}}_g + \mathbf{u}'_g, \tag{5.9}$$

with $\bar{\mathbf{u}}$ the mean (temporal average) velocity and $\mathbf{u}'$ the turbulent velocity fluctuation. Assuming isotropic turbulence, the value of the fluctuating component is distributed normal to the mean fluctuation $\bar{\mathbf{u}} = 0$ m/s and the standard deviation $\sigma_T = \sqrt{2k/3}$, where k is the turbulent kinetic energy. By applying a random number generator, an instantaneous, turbulent kinetic-energy-related value of the fluctuating gas velocity component is calculated and added to the mean gas velocity value. The modified gas velocity still holds until the particle is within a single turbulent vortex. Figure 5.6 shows, schematically, the movement of a particle through a plane flow field characterized by a series of turbulent eddies. The interaction time τ_v between particle and eddy is characterized by the minimum particle passing time through the eddy τ_u and the eddy lifetime τ_T:

$$\tau_v = \min(\tau_u, \tau_T). \tag{5.10}$$

The eddy lifetime (dissipation time scale) is derived from:

$$\tau_T = c_T \frac{k}{\varepsilon}, \tag{5.11}$$

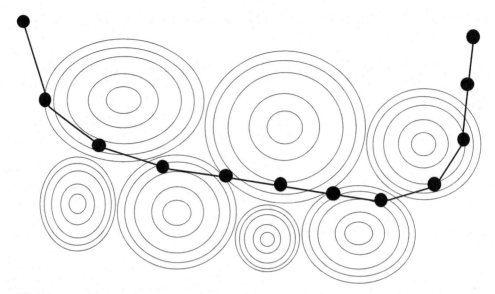

Fig. 5.6 Movement of a particle through a turbulent flow field, idealized as a series of turbulent eddies (Bergmann, 2000)

where the constant c_T has a value of 0.3 (Sommerfeld, 1996). The passing time of a particle through an eddy is calculated from the dissipation length scale L_T and the mean relative velocity:

$$\tau_u = \frac{L_T}{|\mathbf{u}_g - \mathbf{u}_p|},$$

(5.12)

where L_T is calculated from the turbulent kinetic energy distribution as:

$$L_T = \tau_T \sqrt{\frac{2}{3}k}.$$

(5.13)

Within an isotropic turbulent flow field, the three root mean square (RMS) components of gas velocity fluctuations are equal in all three space dimensions $\sqrt{\bar{u}_g'^2} = \sqrt{\bar{v}_g'^2} = \sqrt{\bar{w}_g'^2}$. For axisymmetric (two-dimensional) flow calculation in cylindrical coordinates, fluctuation of the circumferential gas velocity w_g' will always result in an increase of the radial velocity component v_g' by the value of $v_{g,w}'$, as can be seen in Figure 5.7. Therefore, the third velocity component within a two-dimensional spray calculation must be handled separately. The corrected fluctuation value of the gas velocity in a radial direction $v_{g,\text{res}}'$ may be calculated from:

$$v_{g,\text{res}}' = v_g' + v_{g,w}' = v_g' + \left[\sqrt{\left(\frac{r}{\Delta t}\right)^2 + w_g'^2} - \frac{r}{\Delta t} \right],$$

(5.14)

with r the radial position and Δt the time step for iterative calculation of the particle trajectory.

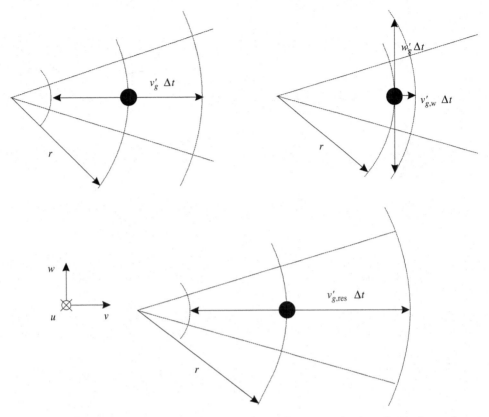

Fig. 5.7 Correction of fluctuating velocity component for calculation in cylindrical coordinate systems (Bergmann, 2000)

As has been pointed out, for example, by Sommerfeld (1996) this model does not account for the effects of crossing trajectories. The turbulent velocity fluctuations, which a gas element along its way and a particle along the particle trajectory within a time step Δt will see, may differ due to particle inertia and external forces acting on the particle, as illustrated in Figure 5.8. The temporal correlation that a fluid element will see may be described at two successive increments by the Lagrangian time correlation coefficient:

$$R_L(\Delta t) = \frac{\overline{u'(t)u'(t + \Delta t)}}{\sqrt{u'^2(t)}\sqrt{u'^2(t + \Delta t)}}, \tag{5.15}$$

which may be approximated in different modelling ways (Sommerfeld, 1996).

5.1.2   Heat transfer and particle cooling

The heat transfer process across the surface of a moving spherical particle in a gas flow field is determined by convection and radiation. Radiation may occur between the particle under investigation and the surrounding spray chamber walls as well as between the particles.

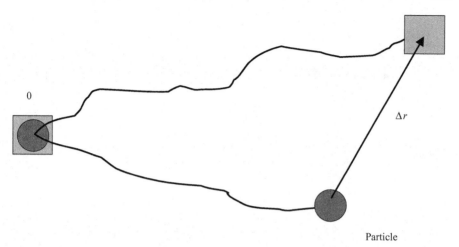

Fig. 5.8 Different flow paths of a particle and a fluid element initially at identical positions (Bergmann, 2000)

The latter is usually neglected. Therefore, the heat balance for an individual particle can be expressed as:

$$m_p c_p \frac{dT_p}{dt} = \mathrm{Nu} \lambda_g \pi d_p (T_g - T_p) - \sigma_S \varepsilon_S (T_p^4 - T_w^4) \pi d_d^2. \tag{5.16}$$

Within a typical spray forming application with twin-fluid atomization of the melt, comparison of the heat fluxes resulting from convection and radiation obtains a heat loss due to convection which is two orders of magnitude higher than the heat loss due to radiation. This effect is due to the huge velocity gradients between gas and droplets and the resulting high heat transfer coefficients. Therefore, in some spray forming investigations, the heat loss due to radiation from the particle has been totally neglected. In the above heat balance equation, the Nusselt number Nu for convective heat transfer is typically taken from the conventional Ranz and Marshall (1952) correlation:

$$\mathrm{Nu} = 2 + 0.6 \mathrm{Re}^{0.5} \mathrm{Pr}^{0.33}. \tag{5.17}$$

An extension of this correlation that takes into account gas turbulence effects during heat and mass transfer from droplets has been derived by Yearling and Gould (1995):

$$\mathrm{Nu} = 2 + 0.584 \mathrm{Re}^{0.5} \mathrm{Pr}^{0.33} \left(1 + 0.34 \sigma_t^{0.843}\right). \tag{5.18}$$

These correlations depend on the local Reynolds number Re, the Prandtl number Pr and the later correlation with the relative turbulence intensity σ_t. The multiplication factor in the brackets of (5.18) therefore extends the conventional Ranz–Marshall correlation.

Solidification modelling

For evaluation of the temporal temperature distribution of particles that (partly) solidify during the flight phase in the spray, specific solidification and phase change models tailored to the material and the process under investigation need to be developed. These models describe the phase change behaviour of a material element from the molten state to the fully solidified state (from the superheating temperature of the melt to the room temperature of the solid particle).

The solidification behaviour of a metal melt is conventionally described by the processes of nucleation and crystal growth. In a first attempt, the solidification process is described by a homogeneous nucleation model without undercooling (equilibrium solidification), which is valid for pure metals at low cooling rates. Here it is assumed that the superheated melt droplet first cools to the phase change temperature (solidification temperature). During the phase change, the latent heat from the droplet is spontaneously released and simultaneously transferred across the surface of the droplet. After solidification has finished, the particle mass cools down further. Within the heterogeneous nucleation model it is assumed that the crystallization process initiates at foreign nuclei. In this model, it is assumed that upon reaching the solidification temperature, a balance occurs between the released latent heat and the heat convectively transferred across the surface of the droplets. The temperature of the droplet remains constant during solidification. These solidification models have been used in spray forming, for example, by Zhang (1994) and Liu (1990).

A more realistic solidification model for metal melts is more complicated to describe. These models may be derived from equilibrium phase diagrams for slow solidifications and time-transfer phase change diagrams and/or from experimental solidification investigation of some realistic cooling rates found in spray processes.

Solidification model in spray forming

The solidification model described here (Bergmann, 2000) has been developed for low-carbon steel C30 (0.3 wt. % C), but may be easily adapted to other material compositions. Figure 5.9 shows part of the iron–carbon phase diagram, where the area for C30 is highlighted. For low cooling rates in equilibrium, the temperature versus time curve shown can be derived directly from this phase diagram. In fact, in spray processes the cooling rate of droplets immediately after atomization may be very high. Therefore, the possibility of undercooling prior to nucleation and the onset of solidification has to be considered. In Figure 5.10 the typical qualitative temperature distribution for a single droplet in a metal droplet spray is shown. Starting with the initial melt temperature (superheated) T_m, the droplet cools down to liquidus temperature T_l. Depending on the actual cooling rate, the droplet may undercool until it reaches the nucleation temperature T_n before solidification starts. Due to the rapid release of latent heat of fusion during recalescence, the droplet temperature increases until it reaches a local maximum in the cooling curve at T_r. During the following segregated solidification, the droplet temperature decreases continuously. At temperature T_{per}, peritectic transformation takes place at constant droplet temperature. After termination of the peritectic transformation, again segregated solidification occurs until the droplet is completely solidified at T_s. From here on, cooling of the droplet

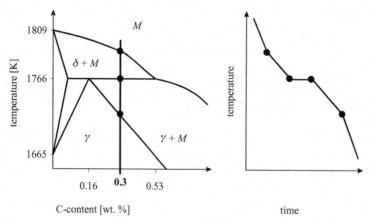

Fig. 5.9 Part of the iron–carbon equilibrium phase diagram (C30 composition is highlighted)

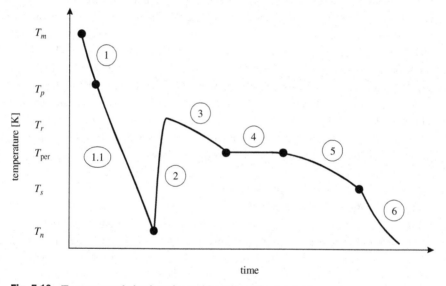

Fig. 5.10 Temperature behaviour for rapid cooling of a droplet in a spray

occurs in the fully solidified particle. In the following, the different states of in-flight droplet cooling and solidification, shown in Figure 5.10, will be analysed separately.

- Cooling in the liquid state (1)
  For a spherical droplet, the change in internal heat content due to convective and radiative heat transfer can be expressed by:

$$c_{d,l} \frac{dT_d}{dt} = -\frac{6h}{\rho_d d_d}(T_d - T_g) - \frac{6\varepsilon\sigma}{\rho_d d_d}(T_d^4 - T_w^4), \tag{5.19}$$

where T_d is the droplet temperature, T_g the gas temperature and T_w the temperature of the surrounding walls. The specific heat capacity of the liquid droplet material is $c_{d,l}$, h is the heat transfer coefficient, ε and σ are the emissivity and Stefan–Boltzmann constant, ρ_d and d_d are the droplet's density and diameter, respectively. A temperature gradient inside the droplet is neglectable, because of the high thermal conductivity of metals, and therefore very low Biot numbers (Bi $\ll$ 1) for all metal droplets can be considered.

- Undercooling (1.1)

When the droplet temperature reaches the liquidus of the material, the solidification process does not immediately start. Depending on cooling rate and droplet size, the temperature T_n where nucleation occurs can be much lower than the liquidus temperature T_l. The nucleation temperature for continuous cooling is defined as the certain temperature, where the number of nuclei N_n in the droplet volume V_d is identical to one:

$$N_n = V_d \int_{T_l}^{T_n} \frac{J(T)}{\dot{T}} dT = 1. \tag{5.20}$$

Here, $J(T)$ is the nucleation rate and $\dot{T}$ the cooling rate (Lavernia, 1996; Lavernia et al., 1996; Pryds and Hattel, 1997). Hirth (1978) has shown that (5.20) may be simplified to:

$$\frac{0.01 J(T_n) V_d \Delta T_{hom}}{\dot{T}} \approx 1, \tag{5.21}$$

where ΔT_{hom} is the undercooling temperature difference for homogeneous nucleation. The nucleation rate may be expressed (Hirth, 1978; Libera et al., 1991) as:

$$J(T_n) = K \exp\left(-\frac{16\pi\sigma_{sl}^2 V_m^2 T_l^2}{3k T_n \Delta h_{fm}^2 \Delta T_{hom}^2}\right), \tag{5.22}$$

with $\sigma_{s,l}$ the solid–liquid interfacial energy and Δh_{fm} the molar latent heat of fusion. From measurements, the pre-exponential factor K is derived as 10^{41} m^3/s^2 (Hirth, 1978; Libera, 1991). In the work of Turnbull (1950) and Woodruff (1973), correlation between the solid–liquid interfacial energy, the latent heat of fusion per atom $\Delta h_{f,a}$ and the atomic volume V_a is given by:

$$\sigma_{s,l} = 0.45 \Delta h_{f,a} V_a^{-2/3}. \tag{5.23}$$

It is well known that in technical processes, heterogeneous nucleation rather than homogeneous nucleation mechanisms limit the degree of undercooling. Only in very small droplets does homogeneous nucleation play an important role during solidification. Based on experimental results for different alloys, Mathur et al. (1989a) derived the following exponential correlation between actual undercooling ΔT and the amount of undercooling necessary for homogeneous nucleation, which can be formulated (Libera et al., 1991) as:

$$\Delta T = \Delta T_{hom} e^{(-2.210^{12} V_d)}. \tag{5.24}$$

Once the actual undercooling is calculated based on the previous set of equations, the actual nucleation temperature for a droplet is determined by:

$$T_n = T_l - \Delta T. \tag{5.25}$$

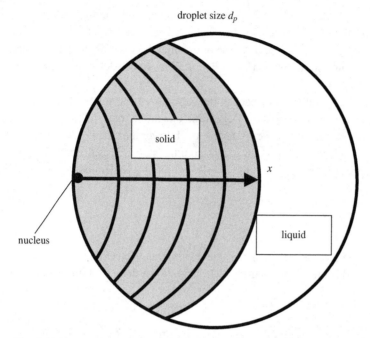

droplet size d_p

solid

x

nucleus

liquid

Fig. 5.11 Movement of the solidification front in a melt droplet from a single nucleus

In this model the maximum value of undercooling is limited based on the results of Turnbull (1950), e.g. the maximum undercooling for iron-based alloys is 295 K. Also a minimum undercooling of 3 K is assumed.

- Recalescence (2)

    After nucleation has begun, the solidification process of a droplet obtains an internal heat source due to release of the latent heat of fusion. The conservation equation for droplet thermal energy has to be extended with a corresponding term to:

$$c_d \frac{dT_d}{dt} = \Delta h_f \frac{df_s}{dt} - \frac{6h}{\rho_d d_d}(T_d - T_g) - \frac{6\varepsilon\sigma}{\rho_d d_d}\left(T_d^4 - T_w^4\right), \tag{5.26}$$

with f_s the solid fraction ($f_s = 0$, droplet is completely liquid; $f_s = 1$, droplet is completely solid) and the specific heat capacity of the droplet c_d as the average of the solid and liquid content:

$$c_d = f_s c_{d,s} + (1 - f_s) c_{d,l}. \tag{5.27}$$

The solidification kinetics in (5.26) may be transformed into the following expression:

$$\frac{df_s}{dt} = \frac{df_s}{dx} \frac{dx}{dt}. \tag{5.28}$$

Assuming that a single nucleation event at the surface of the droplet starts the solidification process and the curvature of the solid–liquid interface during recalescence is equal to droplet surface curvature, as illustrated in Figure 5.11, the change of the solid fraction along the growth axis is given (Lee and Ahn, 1994) by:

$$\frac{\mathrm{d}f_s}{\mathrm{d}x} = \left[\frac{3}{2}\left(\frac{x}{d_p}\right)^2 - \frac{1}{2}\left(\frac{x}{d_p}\right)^3\right]' = \frac{1}{d_p}\left[3\left(\frac{x}{d_p}\right) - \frac{3}{2}\left(\frac{x}{d_p}\right)^2\right]. \tag{5.29}$$

The velocity of the solid–liquid interface movement is approximated to the linear crystal growth rate function of undercooling:

$$\frac{\mathrm{d}x}{\mathrm{d}t} = K_{s,l}[T(f_s) - T_p] = K_{s,l}\Delta T. \tag{5.30}$$

In this equation, $K_{s,l}$ is the solid–liquid interfacial mobility, having a magnitude of 0.01 m/s/K (Lavernia, 1996; Lavernia et al., 1996; Lee and Ahn, 1994; Wang and Matthys, 1992). The phase of recalescence ends when the rate of internal heat production equals the heat transfer from the droplet surface. Here, the cooling curve of a droplet reaches a local maximum (Figure 5.10) and the droplet temperature equals T_r:

$$\Delta h_f \frac{\mathrm{d}f_s}{\mathrm{d}t} = \frac{6h}{\rho_d d_d}(T_r - T_g) + \frac{6\varepsilon\sigma}{\rho_d d_d}(T_r^4 - T_w^4). \tag{5.31}$$

• Segregated solidification 1 (3)
  Further solidification after recalescence again takes place with a decrease in droplet temperature. The heat conservation equation in this stage is described by:

$$\frac{\mathrm{d}T_d}{\mathrm{d}t}\left(c_d + \Delta h_f \frac{\mathrm{d}f_s}{\mathrm{d}T_d}\right) = -\frac{6h}{\rho_d d_d}(T_d - T_g) - \frac{6\varepsilon\sigma}{\rho_d d_d}(T_d^4 - T_w^4). \tag{5.32}$$

The increase of solid fraction with droplet temperature is assumed according to Scheil's equation (Brody and Flemings, 1966):

$$c_s^* = \mathrm{ke}\, c_0(1 - f_s)^{\mathrm{ke}-1} \quad \text{with} \quad c_s^* = \mathrm{ke}\, c_l, \tag{5.33}$$

where c_s^* is the composition of solid at the solid–liquid interface, c_0 is the initial composition of the material and ke is the equilibrium partition ratio. This relation can be transformed into:

$$f_s = 1 - (1 - f_{s,r})\left(\frac{c_l}{c_0}\right)^{\frac{1}{\mathrm{ke}-1}} = 1 - (1 - f_{s,r})\left(\frac{T_{\mathrm{Fe}} - T_d}{T_{\mathrm{Fe}} - T_l}\right)^{\frac{1}{\mathrm{ke}-1}} \tag{5.34}$$

and

$$\frac{\mathrm{d}f_s}{\mathrm{d}T_p} = \frac{1 - f_{s,r}}{(\mathrm{ke} - 1)(T_{\mathrm{Fe}} - T_{p,r})}\left(\frac{T_{\mathrm{Fe}} - T_p}{T_{\mathrm{Fe}} - T_{p,r}}\right)^{\frac{2+\mathrm{ke}}{\mathrm{ke}-1}}, \tag{5.35}$$

with T_{Fe} the liquidus temperature of the pure iron base material, and $T_{d,r}$ and $f_{s,r}$ the solid fraction and temperature of the droplet after recalescence, respectively.

• Peritectic transformation (4)
  When the droplet temperature reaches the peritectic temperature, it remains at a constant value until this phase transformation is completely terminated. The change in solid fraction during peritectic solidification is described by:

$$\Delta h_f \frac{\mathrm{d}f_s}{\mathrm{d}t} = -\frac{6h}{\rho_d d_d}(T_d - T_g) - \frac{6\varepsilon\sigma}{\rho_d d_d}(T_d^4 - T_w^4). \tag{5.36}$$

Peritectic solidification ends when the composition of the remaining liquid reaches the appropriate concentration. Based on phase diagrams, it is possible to calculate the solid fraction $f_{s,pe}$ according to this concentration:

$$f_{s,pe} = \frac{0.53 - c_0}{0.53 - 0.16} = 0.622 \quad \text{(for } c_0 = 0.3 \text{ wt. \%)}. \tag{5.37}$$

- Segregated solidification 2 (5)
  Further segregated solidification takes place in the droplet after peritectic transformation and can be described by the assumptions shown in Phase (3).
- Cooling in the solid state (6)
  After the droplet is completely solidified, it cools down further in the solid state. This process can be evaluated from:

$$c_{d,s} \frac{dT_d}{dt} = -\frac{6h}{\rho_d d_d}(T_d - T_g) - \frac{6\varepsilon\sigma}{\rho_d d_d}(T_d^4 - T_w^4), \tag{5.38}$$

with $c_{d,s}$ the specific heat capacity of the solid material. Cooling in the solid state is interpreted as a submodel of the numerical simulation model of the spray behaviour, taking into account the two-way coupling of momentum and heat.

Model results

Based on the above discussion of quasi-two-dimensional model (Zhang, 1994) calculations (for particle velocity at the spray centre-line during atomization of a steel melt), the results for particle temperatures in flight will now be discussed. These are first based on a heterogeneous phase change model without undercooling. For the boundary condition of gas temperature distribution at the spray centre-line, it is assumed that the gas temperature increases linearly with nozzle distance. This assumption is somewhat arbitrary due to the limited (experimental or numerical) values available here. The final value of the gas temperature distribution has been assumed from measurement of typical exhaust gas temperatures during spray forming processes. A heterogeneous phase change model is assumed for melt droplets without undercooling.

Figure 5.12 illustrates the calculated starting velocity of the gas, i.e. 150 m/s (corresponding to an atomizer gas pressure of 0.15 MPa); and in Figure 5.13 the result for an initial gas velocity of 250 m/s (corresponding to a gas prepressure of 0.4 MPa) is shown. For a lower initial gas velocity value and gas prepressure all particles with sizes $d_p < 200$ μm are calculated to impinge onto the substrate in a fully solidified (solid) state. The remaining liquid for successful compaction during spraying and building of the preform is delivered from bigger droplets. At increased pressure, the diameter limit for solidification shifts towards smaller values. All droplets with $d_p < 100$ μm impinge within 800 mm of the nozzle in a fully solidified state. The remaining thermal energy content of droplets of identical size increases for increasing gas pressure as a result of the decreased flight time in the spray (assuming identical gas temperature and density distributions as in this uncoupled simulation). All bigger particles are in a state of phase change at the point of impingement, and contain the remaining liquid content. Superheating is stopped, in all cases, a short distance

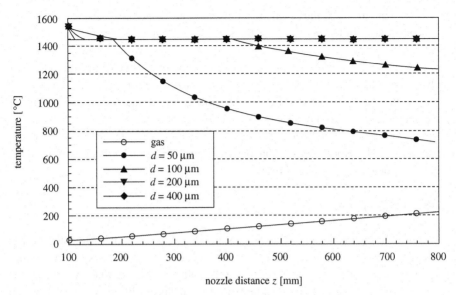

Fig. 5.12   Gas and particle temperature distribution at $v_{g0} = 150$ m/s (Zhang, 1994)

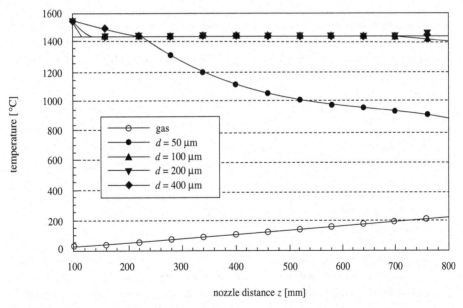

Fig. 5.13   Gas and particle temperature distribution at $v_{g0} = 250$ m/s (Zhang, 1994)

of approximately 100 mm from the atomization area, and this is where the droplets start to change phase. In the main part of the spray, the droplets are in a phase of liquid to solid change.

Up to this point, only the behaviour of the droplets at the centre-line of the spray has been derived. In order to extend these one-dimensional results to a quasi-two-dimensional, the results of the former need to be extrapolated in the radial direction of the spray. The underlying assumption for this extrapolation is based on the local particle size distribution and its measured radial behaviour (i.e. by collecting particles from the spray at relevant positions). Therefore, a log-normal particle size distribution has been assumed and its mean mass value can be calculated from:

$$d_{50.3}(r) = d_{50.3}^{r=0}\left(1 - 0.3\frac{r}{R_0}\right), \tag{5.39}$$

where R_0 is the maximum radius of the spray at the point of impingement (here 10 cm) and $d_{50.3}^{r=0}$ is the mass median particle size at the centre-line of the spray. The median particle size decreases linearly from the centre of the spray to its edge, where the value at the spray edge is 70% of the maximum value within the centre of the spray. In this way, the local value of the integral thermal state of the spray mass at the point of impact is determined. The local-averaged total enthalpy H is calculated from the individual droplet enthalpy:

$$\overline{H} = \sum_i q_3(d)H_i\Delta d_i, \tag{5.40}$$

assuming a finite number of drop-size classes in the total size distribution with particle-size width Δd. The volume (or mass for assumed constant material densities) density distribution of particle sizes is $q_3(d)$ and the summation is performed locally for all particle size classes. The local total enthalpy of each particle-size class is:

$$H = c_p\frac{T - T_{s,l}}{L_h} + f_l, \tag{5.41}$$

where L_h is the melt enthalpy and f_l is the liquid content of each particle size. From this definition of total enthalpy, the solidification state of a droplet is calculated:

- solid (solidified), if $H \leq 0$;
- in the state of phase change (partly solidified), when $0 < H < 1$;
- fluid (fully molten), if $H \geq 1$.

The liquid content of a particle is described by $f_l = 0$ for a fully solidified particle, and by $f_l = 1$ for a fully molten droplet. The corresponding solid content of a particle is $f_s = 1 - f_l$.

The local enthalpy flux entering the top of the sprayed deposit, as a necessary boundary condition for the following calculations, must reflect the local mass flux distribution in the spray as well. It is calculated from:

$$\dot{H}(r) = H\dot{m}. \tag{5.42}$$

The mass flux distribution in the spray to be prescribed here has been derived from

Table 5.3 *Parameters for calculation of one-dimensional particle cooling behaviour*

	T_{p0} [°C]	σ_d [-]	$d_{50.3}^{r=0}$ [μm]	$T_{g,\max}$ [°C]	u_{g0} [m/s]
1	1540	0.6	200	300	150
2	1540	0.4	200	300	150
3	1580	0.4	200	300	150
4	1540	0.4	200	500	150
5	1540	0.4	250	300	150
6	1580	0.4	250	300	150
7	1540	0.4	200	300	250

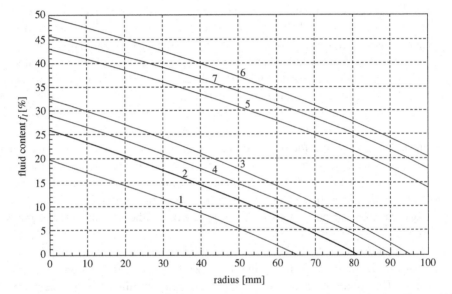

Fig. 5.14 Locally averaged liquid contents in the spray at the spray impact point (Zhang, 1994)

multidimensional simulation of a spray and from experimental measurements (as in Uhlen-winkel (1992)). This will be discussed in the next section.

Below, the mean enthalpy H of the total liquid contents of the impinging spray mass is discussed. The radial distribution at the point of impingement is illustrated in Figure 5.14. The basic parameters used for the simulation runs, for the particle start temperature T_{p0}, the standard deviation of the particle size distribution in the spray σ_d, the mass median particle size $d_{50.3}^{r=0}$ at the spray centre, the maximum gas temperature $T_{g,\max}$ and the initial gas velocity in the atomization area u_{g0}, are listed in Table 5.3.

The basic parameters and boundary conditions for the simulation are shown by Case 2 in the list. The other cases are variations of this basic case. For all, the remaining liquid content of the spray droplets at the impingement distance $z = 800$ mm decreases with increasing radius. In some cases one can observe local liquid contents of 0% at the outer edge of the spray: these results are due to averaging the enthalpy at each location over the whole droplet-size spectrum. This result does not mean that all particles are fully solidified (which will prevent the droplet mass from compaction on the subrate/deposit). The bigger droplets will still contain some liquid (see above section), which will contribute to sticking of the droplets on the surface and to homogeneous preform production. But due to the greater amount of small cold particles, the liquid contribution of large particles to the integrated spray property is not reflected in the averaged quantity (see Section 5.2 on different averaging methods).

In comparison to Case 2, the local liquid content in the spray f_l decreases when the drop-size distribution in the spray becomes narrower, i.e. the standard deviation of the drop-size distribution decreases. This can be seen in Region 1 of Figure 5.10. This scattering parameter of the drop-size distribution cannot be controlled directly within the spray forming process. Here, a change in process operational parameters will result in a simultaneous change in several spray parameters and also in the drop-size distribution. Therefore, changing the standard deviation of the drop-size distribution is a somewhat artificial action. But here a major advantage of simulation models can also be seen. In a simulation run, the results of specific parameter variations may be observed, where some individual parameters are changed, which cannot be independently realized in experiments. Here the potential and behaviour of all physical parameters, boundary conditions and process variables, may be analysed. Decreasing liquid content in the spray is, in this case, attributed to non-linear correlation between the thermal state of individual impinging droplets and the droplet diameter or size. While for bigger droplets the gradient of the correlation is relatively smooth (because bigger particles change their thermal state more slowly due to their larger mass), for smaller droplets a steeper gradient is seen due to a severer dependency. Therefore, when changing the relative width of the particle-size distribution, changes at the lower end of the particle-size distribution for smaller droplets will significantly alter the integral thermal state of the spray.

The liquid content f_l at the point of spray impingement reacts sensibly, especially to changes in the mean drop size of the drop-size distribution in the spray (Case 5 versus Case 2). The content of the remaining liquid increases significantly with increasing mean drop size because less heat is transferred via the drop surface when the total surface decreases due to increasing mean drop sizes, although the resistance, or flight time, of each particle in the spray increases from fragmentation to impingement because of overall lower particle velocities at increasing particle sizes. Also, the initial droplet velocity (or vice versa, the atomization gas pressure) has a significant influence on the remaining liquid content of the spray (Case 7 versus Case 2). For increasing initial gas velocities and gas pressures, from calculation, the liquid content is also expected to increase. This result does not reflect experimental observations within spray forming processes. During experimentation the enthalpy content of the spray decreases with increasing gas pressure, and the spray becomes colder. This computational mismatch reflects a major problem of the uncoupled simulations presented in this chapter. An increase of atomization gas pressure will result in increasing gas velocities and in decreasing droplet sizes in the spray. Therefore, the different process conditions experienced with variable atomizer gas pressure are best evaluated by comparing Case 6 (for lower atomizer gas pressure) and Case 7 (for increased atomizer gas pressure). This comparison shows that the remaining liquid content in the impinging spray, in agreement with experimental results, decreases. But in this case, the assumed values for gas velocity and mean drop size in the spray cannot be directly correlated with a specific gas pressure. Therefore, the comparison only verifies the trend.

Somewhat more disturbing with respect to the mismatch between simulated results and experimental findings are the lack of coupling of the particulate phase to gas phase behaviour and neglect of the resulting momentum sink within the gas phase. Kramer *et al.* (1997) also found this effect in their two-dimensional calculations of particle behaviour in the spray during spray forming without taking into account coupling effects. They assumed an empirical correlation for gas velocity distribution and gas temperature distribution. Their results also indicate a slightly increasing liquid content in the spray for increasing gas mass flow rates (and therefore increasing gas pressures).

Solidification behaviour inside the melt particle

The temperature distribution inside a single spherical particle during solidification has been studied numerically by Kallien (1988) and Hartmann (1990). The simulation program used has been developed from an original code for simulation of solidification during metal casting. It calculates the three-dimensional temperature distribution in spherical particles, and includes possible undercooling. The model is based on solution of the Fourier differential equation for transient heat conduction in three-plane (Cartesian) coordinates as:

$$\rho c_p \frac{\partial T}{\partial t} = \frac{\partial}{\partial x}\left(\lambda \frac{\partial T}{\partial x}\right) + \frac{\partial}{\partial y}\left(\lambda \frac{\partial T}{\partial y}\right) + \frac{\partial}{\partial z}\left(\lambda \frac{\partial T}{\partial z}\right), \tag{5.43}$$

where the thermophysical properties: conductivity λ, density ρ and heat capacity c_p, depend on location and temperature. A modified temperature is introduced:

$$\Theta = \frac{1}{\lambda_0}\int_0^T \lambda\,dT \tag{5.44}$$

to achieve a linear differential equation:

$$\frac{\rho c_p}{\lambda_0}\frac{\partial T}{\partial t} = \frac{\partial^2 \Theta}{\partial x^2} + \frac{\partial^2 \Theta}{\partial y^2} + \frac{\partial^2 \Theta}{\partial z^2}. \tag{5.45}$$

This equation is solved by means of a finite difference method on an orthogonal-plane three-dimensional grid. The geometry of the spherical particle is reflected by a stepwise arrangement of cubical cells. A number of 24 000 grid cells has been used for the calculation. The assumed boundary conditions are:

- the surrounding gas atmosphere is assumed to be at constant temperature,
- the heat transfer coefficient is taken as constant across the whole surface of the particle,
- the onset of solidification is initiated at a preselected nucleation temperature – nucleation may be assumed to start either at a single point (specified grid cell), or at a number of grid cells for heterogeneous nucleation modelling, or simultaneously within all grid cells for homogeneous nucleation modelling.

The derived calculation model for particle solidification behaviour assumes a six-stage approach:

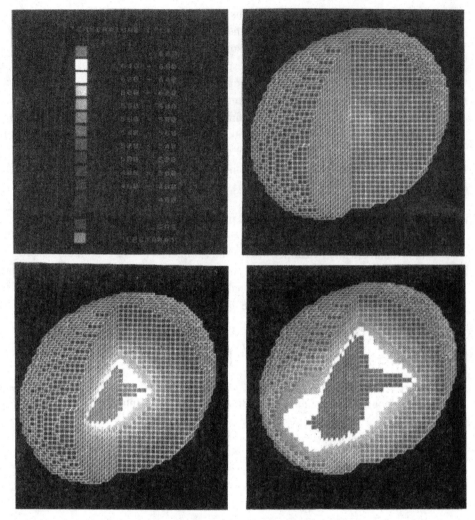

Fig. 5.15 Solidification behaviour inside an aluminium droplet of $d_p = 50\ \mu m$, starting from a single nucleus (in the back of the particle), at 50 K undercooling (Kallien, 1988)

(1) cooling of the melt from superheating until the nucleation temperature is reached,
(2) attaining the highest undercooling,
(3) onset of solidification and recalescence,
(4) cooling and solidification in the melt temperature range between solidus and liquidus,
(5) end of solidification,
(6) cooling of the solidified particle.

For the recalescence phase, it is assumed that the velocity of latent heat release for an aluminium particle under investigation depends on undercooling ΔT:

$$v = K\Delta T. \tag{5.46}$$

The kinetic factor K for aluminium takes values between 2 cm/K s and 2 m/K s. The solidification rate of a grid cell of volume V is:

$$\frac{dV}{dt} = v^* s^2. \tag{5.47}$$

When a grid cell is completely solidified the neighbouring grid cells begin to release their latent heat. If the temperature of these grid cells has already been raised by latent heat conduction from the first solidified element, these grid cells will solidify somewhat slower, as described above. The velocity of the solidification front therefore changes, depending upon the undercooling rate of the elements during solidification.

The result of calculation for an aluminium particle of 50 μm diameter is shown in Figure 5.15. The heat transfer coefficient is assumed to be $\alpha = 20\ 000$ W/m² K and the undercooling prior to nucleation is taken as 50 K. The solidification process initiates at a single point on the surface of the particle in a plane inside the particle. By release of latent heat, the interior of the particle is heated up. For a 10% solidification rate, movement of the solidification front is visible, which raises the temperature of the surrounding grid cells close to the liquidus temperature.

Due to the low Biot numbers typically achieved in melt particles in gaseous atmospheres (high heat conductivities of metals) the thermal behaviour inside a particle is typically not taken into account during proper thermal particle simulation. Without loss of accuracy, the temperature gradient inside the melt particle may be neglected, and a mean constant particle temperature assumed.

This inspection of the thermal behaviour of particles in a spray underline the necessity for refined modelling of spray behaviour in order to obtain realistic spray forming simulation results. Such refinement within a multidimensional, multicoupled spray simulation will be introduced in Section 5.2.

5.2 Internal spray flow field

Multicoupled models of spray processes are necessary in order to derive realistic and useful simulation results. Based on such models, investigation of exchange and transfer processes in two-phase spray flows is possible.

In the following, a multidimensional, multicoupled simulation model for spray processes based on a Eulerian/Lagrangian approach will be introduced. First, conservation equations for mass, momentum and thermal energy, in combination with a turbulence model for the gaseous phase, will be used to investigate the pure gas flow field in a two-dimensional domain, i.e. gas flow in the atomizer and spray chamber. The discretized equations have been solved by means of a finite volume algorithm.

The fundamental conservation and transport equation is given in (3.1). In the case of two-dimensional flow, this equation simplifies for time-invariant flow of a continuous phase to:

$$\frac{\partial}{\partial x}(\rho u_g \Phi) + \frac{1}{r}\frac{\partial}{\partial r}(\rho r v_g \Phi) - \frac{\partial}{\partial x}\left(\Gamma \frac{\partial \Phi}{\partial x}\right) - \frac{1}{r}\frac{\partial}{\partial r}\left(r \Gamma \frac{\partial \Phi}{\partial r}\right) = S_\Phi - S_{\Phi_p}. \tag{5.48}$$

Once again, in this equation Φ is the transport variable and Γ is the effective viscosity. The source terms are defined as S (the index Φ is for internal sources/sinks and the index Φ_p is for sources/sinks from the phase coupling).

Second, the trajectories or flight paths of a great number (several 1000 up to several 10 000, dependent on the flow type) of individual particles (respectively parcels of particles displaying identical properties) are calculated. Interaction of these parcels with the gas phase will be stored at the control volume locations of the solution domain. The source terms in the conservation equations, describing coupling of the dispersed phase to the continuous phase, for phase coupling of momentum and thermal energy, are derived from the particle-source-in-cell approach (Crowe et at., 1977):

$$S_{\text{mom}} = 3\pi \mu_g d_p f_r (u_p - u_g) N, \tag{5.49}$$

$$S_{\text{ener}} = \text{Nu}\pi \lambda_g d_p (T_p - T_g) N. \tag{5.50}$$

In these source terms, f is the normalized friction coefficient as a ratio of particle resistance (for solid particles expressed by means of Eq. (5.4)) to Stokes resistance ($c_d = 24/\text{Re}$ for Re $\ll 1$):

$$f_r = 1 + 0.15\text{Re}^{0.687}. \tag{5.51}$$

The thermal energy source term contains the Nusselt number Nu for convective heat transfer from the particle to the gas (from Eq. (5.16)) as well as the instantaneous local Reynolds number Re, which is expressed in terms of the local relative velocity between phases. The number N is the number of particles in a discrete volume of the solution domain.

The momentum and thermal energy exchange values between particulate and continuous phases will be summed up locally at all grid volumes over all included particle parcels to obtain the source terms for the continuous phase conservation equations.

The third step in the coupled solution is renewed calculation of the continuous phase conservation equations, which now include the source or sink terms for the momentum and energy equation at all control volumes. By doing so, coupled simulation of a multiphase flow problem (two-way coupling) is achieved. Steps 2 and 3 will be iteratively solved until a prescribed convergence criterion is reached and a final solution for the steady state of the spray process has been achieved.

Description of the thermal state of particles in the spray during solidification is based on the appropriate models for nucleation and crystal growth, as discussed above. The models are material dependent and include the thermal cooling condition (cooling rate) as the main boundary condition.

Boundary conditions for the coupled approach

The calculation is performed for the cylindrical spray forming chamber illustrated in Figure 5.16. The chamber diameter is 1.0 m and the chamber height is 0.8 m. Inflow boundary conditions for the gas consist of two separate inflow ports. The central gas jet

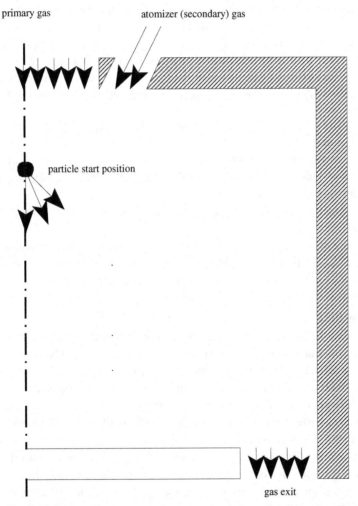

Fig. 5.16 Calculation area for the spray simulation: two-dimensional representation of the spray chamber (Bergmann *et al.*, 1995)

models the primary gas inlet of the free-fall atomizer and is directed vertically downwards. The coaxial slit nozzle represents the secondary flow of the free-fall atomizer in the two-dimensional representation, where the atomization gas exits at a certain inclination angle towards the centre-line. The mass flow rates of the primary and secondary gas are calculated from Eqs. (4.24) and (4.25) for prescribed gas pressures. The gas velocity is equal to the local velocity of sound, thereby the gas flow is assumed to flow in an underexpanded state out of the nozzles. Change of mass flow rates for varying gas pressures is therefore performed by alteration of the gas exit area. The gas exhaust is located, for simplicity, at the outer edge of the lower wall of the spray chamber (slit configuration).

The most important boundary condition for the dispersed phase in the simulation is the starting condition of the droplets at the fragmentation point of the melt. Because the features of the disintegration and atomization processes may not be simulated in detail down to the drop-size scale, physically realistic assumptions of the starting conditions are made. These are based either on simplified fragmentation models, as described in Section 4.3.2, or on empirical input data from experiments. The polydispersed state of particles of different sizes within the spray may, especially, lead to analysis problems. But proper description of this behaviour is important for calculation of the thermal state of the compacting mass of particles. The following assumptions are made in the multicoupled simulation below.

- The starting area for droplets is located a distance 100 mm below the gas exit, reflecting the principal geometry of a free-fall atomizer.
- The particle-size distribution is assumed to be log-normal. The mass median value of the particle-size distribution is calculated from Lubanska's empirical correlation (Eq. (4.49)), and a constant logarithmic standard deviation of $\sigma_{\ln} = 0.7$ is assumed.
- The starting velocity of the droplets equals the calculated melt jet velocity at the fragmentation point (see Section 4.1.2).
- The temperature of the droplets at their starting point is equal to the melt exit temperature at the tundish (see Section 4.1.1); therefore, temperature loss from the melt jet between the exit from the tundish and the disintegration area is neglected.

Single-phase gas flow field

First, proof needs to be obtained as to whether the simulation results of the single-phase gas flow field of the combined atomizer and the primary gas jets in the spray chamber agree with experimental gas velocity measurements in the spray chamber. Figure 5.17 shows measured (Uhlenwinkel, 1992) gas velocity distributions in a spray chamber without atomization for different pure gas pressures. Bergmann et al.'s (1995) simulated results are plotted for comparison at an atomizer gas flow of 2.0 bar. It was not possible to record measurements near the nozzle exit due to geometric restrictions in positioning the measurement probe here. The calculations in this area indicate negative velocity values for an upward pointing gas flow at the centre-line of the jet directly below the primary gas exit. In this region, slowly moving gas from the central nozzle is sucked into the faster moving atomization gas stream. Therefore, recirculation of gas in this area results (see Section 4.2.1). Especially at greater distances from the atomizer, agreement between calculation and measurement is

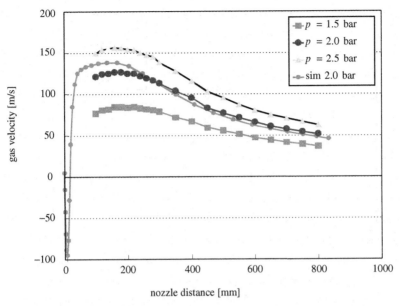

Fig. 5.17 Single-phase gas velocities at the centre-line (Bergmann *et al.*, 1995). Measured data from Uhlenwinkel (1992)

good. Close to the nozzle, the maximum gas velocity achieved is recorded somewhat closer to the atomizer exit area than in the experiment. Also, the simulation overpredicts the value of the maximum gas velocity.

Figures 5.18 and 5.19 illustrate the measured and simulated radial profiles of axial velocity components of the gas flow field in the spray chamber at certain distances from the nozzle. In both descriptions, a local minimum at the centre-line is found, reflecting the geometry of the free-fall atomizer used here (see Section 4.2). At somewhat greater distances from the atomizer, the gas flow field shows a configuration comparable to that of a free turbulent gas jet. The jet spreads radially while the maximum gas flow velocity at the centre-line of the spray decreases. Figure 5.20, showing the distribution of gas velocity and pressure (normalized to the local maximum value within each profile) at the centre-line of the flow field, near the nozzle, exemplifies that a maximum pressure distribution occurs well above the fragmentation area of the liquid. Meanwhile, the maximum velocity distribution at the centre-line is located below the atomization area of the liquid melt in twin-fluid atomization. The pressure maximum is due to reflection of the impinging atomizer gas (stagnation pressure). The role of the pressure distribution in the fragmentation processes in liquid atomization applications is still not clear.

Two-phase flow field

The coupled simulation results of the spray behaviour in spray forming applications now presented are taken from Bergmann *et al.* (1995). These simulations have been performed for spray forming applications of low-alloyed steel materials. The basic parameters and

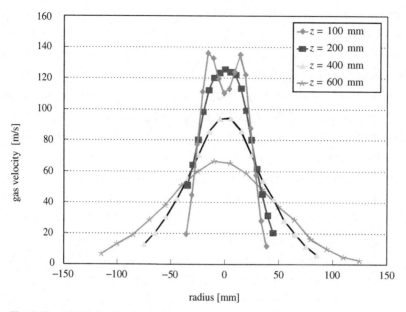

Fig. 5.18   Radial distribution of axial gas velocities (Bergmann *et al.*, 1995)

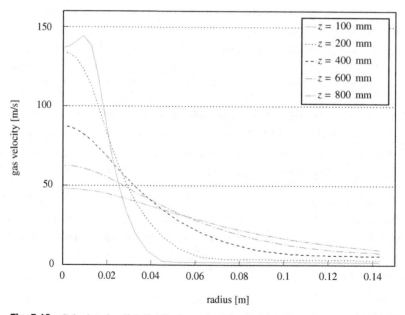

Fig. 5.19   Calculated radial distribution of axial velocities (Bergmann *et al.*, 1995)

Table 5.4 *Parameters for two-phase spray simulation*

p_g [MPa]	$\dot{M}_g$ [kg/h]	$\dot{M}_l$ [kg/h]	1/GMR [-]	ρ_p [kg/m³]	$d_{50.3}$ [μm]
0.2	465	558	1.2	6900	204
0.3	695	417	0.6	6900	126
0.3	695	558	0.8	6900	133
0.3	695	834	1.2	6900	147
0.4	925	558	0.6	6900	104
0.5	1155	558	0.6	6900	88

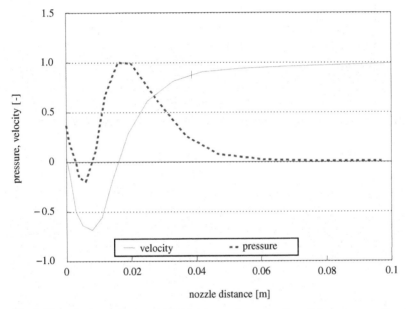

Fig. 5.20 Velocity and pressure distribution at the centre-line of the spray near the atomizer vicinity (Bergmann *et al.*, 1995)

boundary conditions assumed throughout the calculations are listed in Table 5.4. The basic case is shaded; the other five cases are variations from that basic process condition.

The distribution of axial gas velocities and some preselected drop-size classes at the centre-line of the spray is shown in Figure 5.21 for the basic parameter set at an atomization pressure value of 0.3 MPa. The gas flow field above the fragmentation area remains unchanged compared to single-phase flow. In the fragmentation area, a drastic decrease in the velocity of the gas flow is to be seen, which is due to momentum transfer from the gas to the droplets resulting from the drop in acceleration. At somewhat greater distance from the fragmentation area, the gas velocity recovers and increases once again. Here, local expansion of gas due to the rise in temperature associated with release of heat from the droplets occurs (see Figure 5.22), which results in local acceleration of the gas flow. Another reason

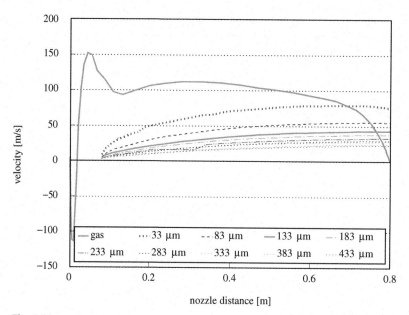

Fig. 5.21   Gas and particle velocities on the spray centre-line (Bergmann *et al.*, 1995)

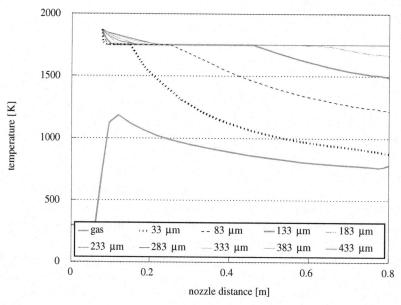

Fig. 5.22   Gas and particle temperatures at the spray centre-line (Bergmann *et al.*, 1995)

for this behaviour is the ongoing entrainment of external gas from the spray chamber across the outer edge of the spray into the spray cone and the mixing with the principal atomizer gas in the centre of the spray. The maximum gas velocity in the case of a coupled two-phase flow simulation is reached at a distance of approximately 330 mm below the nozzle. From this point onwards the gas velocity decreases constantly. Immediately above the lower substrate at the bottom of the spray chamber, the gas rapidly decelerates and is radially reflected towards the exhaust system. The gas velocity in the stagnation point by definition is zero. Reflecting the overall gas velocity distribution in the main part of the spray system (at least within the region at the spray centre-line discussed here) the gas velocity value is almost constant, the change in gas velocity and its decrease at somewhat greater nozzle distances is, by far, less pronounced compared to the case of single-phase gas flow. The distribution of particle velocities at the centre-line of the spray in Figure 5.21 illustrates acceleration values that are much lower compared with values calculated in the uncoupled simulation (see p. 124). The main reason is the low gas density due to rapid heating in the atomization area that is accounted for in the coupled simulation by increased gas temperature. Thereby, the particle resistance force, for acceleration of the droplets, is decreased. The impingement velocities of the particles on the deposit/substrate are calculated well below those values obtained in the simulation for the case without coupling effects taken into account.

The distribution of gas and particle temperatures at the centre-line of the spray for a 0.3 MPa atomization gas pressure is illustrated in Figure 5.22. In this approach, for analysis of the solidification process, a simple heterogeneous nucleation model has been used to describe the phase change of the melt droplets. From the gas temperature distribution, one can see a very rapid rise in the fragmentation area up to values of 1200 K. The maximum gas temperature is reached a few centimetres below the atomization area. During further development of the gas plume, the temperature decreases, mainly because of radial mixing with the entrained colder gas from the outer spray chamber. Because the gas in the total spray area under investigation (at least for that part of the spray discussed here) is somewhat colder than for all calculated droplet classes, the direction of heat transfer is always from the hot droplets to the colder gas. Particle superheating at the point of droplet formation is assumed to be 125 K in this case. After losing their superheat, in a very narrow region, temperature of the droplets immediately remains constant throughout the solidification process in the fragmentation area. This area of droplet phase change covers most of the spray development in this example. From this simulation run, it may be deduced that for the assumed gas mass flow rate (respectively, gas pressure), all particles less than 210 μm in diameter are already fully solidified (solid particles) when they impinge onto the substrate. All bigger particles are in a state of solidification, no particles can be seen which are still fully liquid upon impact.

From analysis of the radial distribution of gas temperatures at specific distances from the atomizer, which are shown in Figure 5.23, three distinct areas in the spray may be identified. The first is the entire 'core region' of the spray. Here the maximum gas temperatures are to be seen. Due to the relatively high particle concentration in this area, the total heat transfer rates (summing up all individual contributions from calculated particles) between the particles and the gas are highest here. At the outer edge of the spray ($r < 0.1$ m), drastically lower

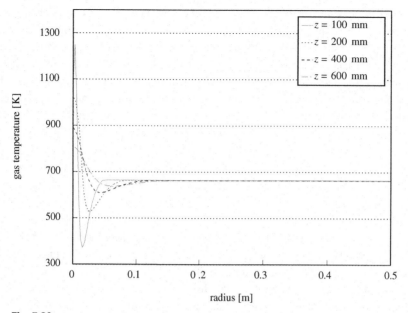

Fig. 5.23 Radial distribution of gas temperature (Bergmann *et al.*, 1995)

gas temperatures are to be seen: this is because the gas exiting from the atomizer nozzles has not been sufficiently warmed up due to the low particle concentration in this area. In the third area, located between the spray and the chamber wall ($r > 0.1$ m), which covers a large part of the geometric extent of the spray in this case, the gas temperature is almost constant. This is the area where recirculation of gas from the lower part upwards into the upper part of the spray chamber occurs, and where it is entrained into the spray plume. In this case, heat transfer between the gas and the spray-chamber wall is not taken into account: the chamber walls are assumed to be adiabatic (this assumption may be easily removed from the simulation). For steady-state conditions, the gas temperature in the recirculation area is approximately 675 K. This seems to be a realistic value as it is well within typically measured gas exhaust temperature levels for steel melt spray forming processes.

The related radial profiles of axial gas velocities at specific distances to the atomizer are illustrated in Figure 5.24. Drastic heating and expansion of the gas in the disintegration area leads to intense radial spreading of the spray cone when compared to similar applications of sprays with cold liquids/droplets without heat transfer. In the central area of the spray, the gas velocities are somewhat smaller than at the spray edge during early stages of spray development. Due to the high particle concentration in the centre of the spray, transference of momentum from the gas towards the droplets is greatest in this region. Only at greater distances from the atomizer will the gas velocity profiles achieve the expected, almost bell-shaped (Gaussian), profile due to radial mixing of the gas. The maximum gas velocity value here is to be seen at the core of the spray. Negative velocity values close to the spray-chamber wall indicate the upward movement of gas from the recirculation area. The radial distance from the centre-line where upward gas movement begins is almost unchanged in the spray

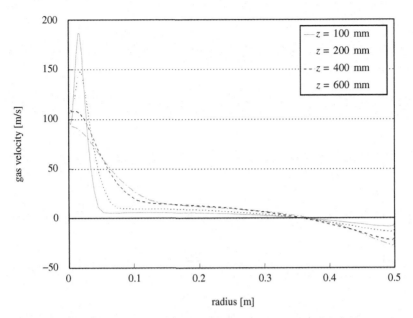

Fig. 5.24   Radial distribution of gas velocity (Bergmann *et al.*, 1995)

chamber, therefore indicating that the radial core of the recirculating vortex is constant with respect to axial distance.

Comparison between two different velocity and temperature trajectories for droplets exhibiting mass median spray particle size is illustrated in Figures 5.25 and 5.26. Here the trajectory of a droplet at the centre of the spray is compared with that at the outer edge of the spray. The analysis must take into account thermal coupling between the continuous and dispersed phases for proper understanding and description of the spray behaviour in spray forming. Because of the somewhat lower gas temperatures at the edge of the spray and the related higher gas densities in this area (when compared to the centre of the spray) higher momentum transfer rates due to increased resistance forces are to be found here. Therefore, the particles at the spray edge are more rapidly accelerated than in the core and move at higher velocities. This behaviour results in shorter flight and residence times for those particles moving at the spray edge than those moving in the spray core. Though the flight time (and therefore the interaction time between gas and droplets) at the spray edge is smaller, the droplet temperature for these edge particles at the point of impingement onto the substrate/deposit is lower than for core particles. The reason for this is the smaller gas temperature at the spray edge and the resulting increased temperature difference due to increasing heat transfer rates of individual particles. This, of course, is only valid for identical particle-size classes: when the integral spray behaviour is calculated in terms of heat fluxes, one has to keep in mind that the local drop-size distribution and the mean drop size in the spray, as well as the mass flux, decrease radially.

By changing the operational parameters of the spray forming process as listed in Table 5.4, the simulated results illustrated in Figures 5.27 and 5.28 are obtained. By alteration

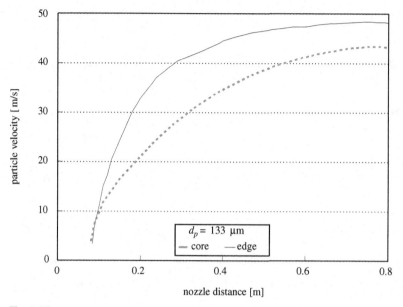

Fig. 5.25 Comparison between particle velocity distributions at the centre-line of the spray and at the spray edge (Bergmann *et al.*, 1995)

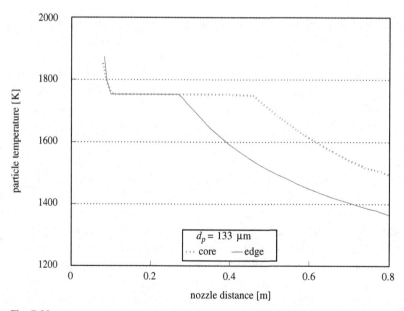

Fig. 5.26 Comparison between particle temperature distributions at the centre-line of the spray and at the spray edge (Bergmann *et al.*, 1995)

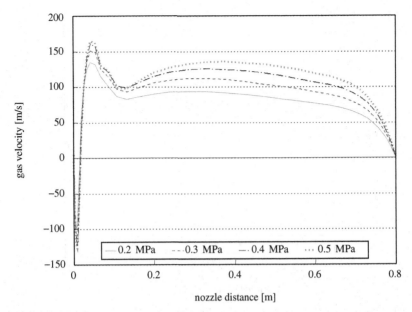

Fig. 5.27 Gas velocity at the spray centre-line for different atomization gas pressures (Bergmann *et al.*, 1995)

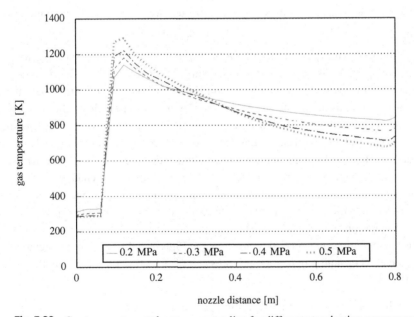

Fig. 5.28 Gas temperature at the spray centre-line for different atomization gas pressures (Bergmann *et al.*, 1995)

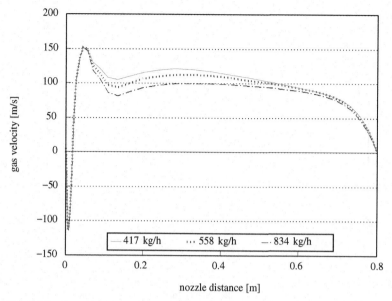

Fig. 5.29 Gas velocity at the spray centre-line for different melt mass flow rates (Bergmann *et al.*, 1995)

of the atomization gas pressure:

(a) increase of the gas velocities in the atomization area occurs, and simultaneously
(b) a decrease of the mean droplet size in the spray results.

In combination with these effects, the heat and momentum transfer rates of an individual mean-sized particle, as well as for the whole particle collection, is increased. But the overall behaviour of the spray is still in question, as the particle concentration increases (at constant mass flow rates of the melt) and the residence or flight time of particles in the spray decreases. As shown in Figure 5.27, in the two-phase flow field in the spray, the gas velocity increases when the gas pressure is raised (as has been shown for the spray centre-line). This is mainly because of intensified heating of the gas in the atomization area, as can be seen in Figure 5.28. At increased nozzle or atomizer distances, the gas temperature changes with increasing atomization pressure. Here the temperature of the gas in the impingement area of the spray decreases due to more intense mixing with the overall colder gas in the spray chamber, for increasing gas pressures.

Gas temperatures and velocities are shown in Figures 5.29 and 5.30 for a changed metal flow rate at a constant atomization gas pressure of 0.3 MPa. By increasing the melt liquid flow rate:

(a) the thermal energy input into the spray is increased, and
(b) the resulting droplet sizes in the atomization spray are augmented.

Therefore, the local gas temperature in the spray (here again shown for the conditions at the spray centre-line only) is, generally, increasing. From analysis and comparison of Figures 5.27 to 5.30 one can observe that classification of the spray forming process on the

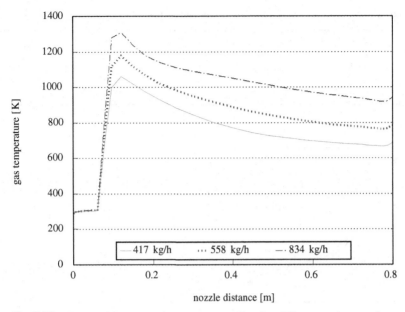

Fig. 5.30 Gas temperature at the spray centre-line for different melt mass flow rates (Bergmann *et al.*, 1995)

basis of the mass flow ratio between gas and melt (GMR) is not sufficient, and, at least, the resulting change in particle-size distribution must be accounted for.

Reliable measurement results for comparison with simulated results, e.g. for droplet velocity or temperature distributions or observation of gas behaviour in the spray, in spray forming applications are rare. Some in-flight measured data from PDA measurements are given in Bauckhage (1998a,b) and Domnick *et al.* (1997) and will be discussed later. Development of suitable methods for detection of droplet properties in the spray during spray forming is an important research topic. For verification of two-phase flow simulation of spray behaviour, often the measured properties of particles collected at the point of spray impingement are taken. Figure 5.31 shows calculated and an experimentally determined results for the radial behaviour of a mass flux of droplets from the centre of the spray to its edge. The simulated values show some deviations, especially in the spray core. This deviation may be due to deficiencies in the model, where the quite crude assumption of droplet starting conditions may be a main factor.

Stochastic contribution to spray structure

An example of results from a fully coupled spray simulation (Bergmann, 2000; Bergmann *et al.*, 2000) is shown in Figure 5.32, where the calculated temperature and solid fraction of individual different size particles at a certain position along the spray centre-line ($r = 0$ m) at a distance $s = 0.27$ m below the point of atomization are to be seen. The process parameters used in this calculation are those characteristic of spray forming of C30 steel, and are shown in Table 5.5. Due to the turbulent character of the flow field and the stochastic nature of particle trajectories through the spray (turbulent droplet dispersion), at a single

Table 5.5 *Parameters for fully coupled spray simulation (Bergmann, 2000)*

Material	Gas pressure p_g [MPa]	Gas mass flow rate $\dot{M}_g$ [kg/s]	Melt mass flow rate $\dot{M}_l$ [kg/s]	GMR [-]	Melt superheat ΔT [K]
Steel C30	0.35	0.21	0.175	1.2	150

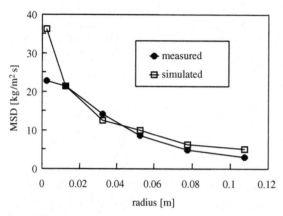

Fig. 5.31 Comparison between measured and calculated mass flux distribution at the spray impingement area: nozzle distance $z = 688$ mm, material: steel C30, GMR $= 1.1$ (Fritsching *et al.*, 1997b)

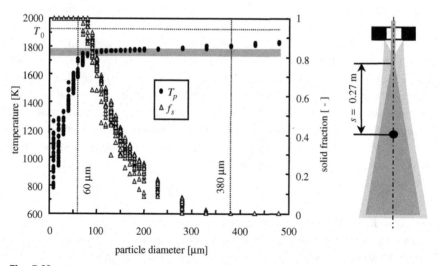

Fig. 5.32 Calculated temperature and solid fraction shown by individual particles at a position $s = 0.27$ m within the spray ($r = 0$ m). The solidification interval of C30 steel is indicated in grey (Bergmann, 2000)

point within the spray (as seen here) the temperature and velocity of identical-sized particles may vary. The following conclusions for the thermal state of the particle with respect to droplet diameter at the observed position in the spray can be deduced from Figure 5.32:

- $d_p < 60$ μm. All particles are already solidified ($f_s = 1$) at temperatures below the solidus temperature. With decreasing particle size, the mean temperature also decreases, while the scattering of individual particle temperatures may vary stochastically from the mean particle class value. Increased scattering of temperatures is due to decreased particle inertia and mass, and to the turbulence in the gas flow, which results in deviation from the deterministic droplet flight path (turbulent dispersion).
- 60 μm $< d_p < 380$ μm. In this droplet-size spectrum, the particles are in the process of solidifying ($0 < f_s < 1$). A scatter may be observed in the thermal state of individual particles: in the 70 to 80 μm size range some fully solidified particles can be seen, while in the 270 to 330 μm size range some still fully fluid droplets can be found. Those particles that are in the process of solidification, show temperatures between the solidus and liquidus temperature (this region is indicated by the grey bar in the temperature plot). The fraction of particles solidifying decreases with increasing particle size.
- 380 μm $< d_p$. All particles within this size spectrum are still fully liquid in the region ($s = 0.27$ m) under consideration. Their temperature is above the liquidus temperature, which means that these droplets still contain some superheat. The amount of superheat increases with increasing particle size.

This scattering of droplet temperature data and solid fractions in the spray means that the temperature of the spray at impingement needs to be suitably averaged. These conditions are due to local droplet-sized distributions and stochastic particle behaviour resulting from turbulent dispersion of droplets in the gas flow fluid.

Thermal averaging of spray conditions

An important condition for successful preform evolution is, besides the mass flux distribution in the spray, derivation of the distribution of the local thermal state of the spray at impingement. Here the local (but particle-size averaged) enthalpy flux in the deposit, as a main measure, can be averaged in two different ways, both of which have been described by Bergmann (2000) and Bergmann *et al.* (1999). The total enthalpy of the impinging droplet mass in the spray can be locally over all drop size classes:

$$\bar{h}_p = \frac{1}{\sum_i m_{p,i}} \sum_i [m_{p,i} \{ [c_l(T_{p,i} - T_s) + \Delta h_f](1 - f_{s,i}) + c_s[(T_{p,i} - T_s)f_{s,i} + T_s] \}$$

(5.52)

Two different cases can be derived:

(1) The first averaging principle calculates the state of solidification and the temperature of the impinging droplets in thermal equilibrium (caloric averaging) dependent on the local state of solidification. Three subcases can be defined:

(1a) The droplet material is already totally solidified ($h_p \leq c_s T_s$):

$$\overline{f}_s = 1, \tag{5.53}$$

$$\overline{T}_p = \frac{\overline{h}_p}{c_s}. \tag{5.54}$$

(1b) The droplet material is in the stage of solidification or is partly solidified ($c_s T_s < c_l(T_l - T_s) + \Delta h_f$):

$$\overline{f}_s = \frac{c_s T_s + c_l(\overline{T}_p - T_s) + \Delta h_f - \overline{h}_p}{(c_l - c_s)(\overline{T}_p - T_s) + \Delta h_f}, \tag{5.55}$$

$$T_h = \frac{\overline{h}_p + (1 - \overline{f}_s)[(c_l - c_s)T_s - \Delta h_f]}{(1 - \overline{f}_s)c_p + \overline{f}_s c_s}. \tag{5.56}$$

The values of f_s and T_h are based on thermal equilibrium assumptions with the aid of an appropriate phase diagram, for alloys; for pure metals, $T_h = T_{s,l}$.

(1c) The droplet material is totally liquid ($c_s T_s + c_l(T_l - T_s) + \Delta h_f < h_p$):

$$\overline{f}_s = 0, \tag{5.57}$$

$$T_h = \frac{\overline{h}_p + (c_l - c_s)T_s - \Delta h_f}{c_l}. \tag{5.58}$$

(2) The second averaging method is called the separation method. It is based on the assumption that the droplets are in an unbalanced state with respect to the solidification process. In this method, energy exchange within the droplet mass is due to its thermal state and is not a result of the remaining solidification enthalpy which is contained in individual droplets:

$$f_s = \frac{\Sigma_i(m_{p,i} f_{s,i})}{\Sigma_i m_{p,i}}. \tag{5.59}$$

$$T_m = \frac{1}{[c_s \overline{f}_s + c_l(1 - \overline{f}_s)]\sum_i m_{p,i}} \sum_i \{m_{p,i} T_{p,i}[c_s \overline{f}_s + c_l(1 - \overline{f}_s)]\}. \tag{5.60}$$

These two averaging methods represent two extreme cases. The first method describes the thermal state of the particulate mass in thermal equilibrium. This means that the thermal state is characterized by means of the specific-enthalpy-related thermal state with respect to temperature and degree of solidification. As discussed earlier, this averaging method may lead to the possible state that the spray material is fully solidified, on average; though in the total droplet mass, some liquid may still be contained in the bigger droplets. This is the case where a huge number (or better mass fractions) of small solidified and relatively cold particles exist in the spray. When averaging by Method 1, the fact that some bigger droplets may still contain liquid will not be reflected. Based on the separation method, this deficit may be avoided. This method yields the amount of liquid, respectively solidified mass, as well as the amount of remaining solidification enthalpy within the spray.

Results of different thermal averaging methods

Figure 5.33 shows the calculated temperature and solid fraction values of particles from Figure 5.32. In addition, the figure shows the averaged values above ($s = 0.04$ m) and below ($s = 0.62$ m) this position in the spray cone. From these diagrams it is obvious that the limits for totally solidified and totally liquid particles are shifted towards larger particle sizes with increasing flight distance. Also the size spectrum of the particles in the state of solidification becomes broader. At larger distances from the point at which particles start ($s = 0.27$ m and $s = 0.62$ m), no significant differences between the two averaging methods is observed. But at short particle-flight distances ($s = 0.04$ m), both averaging methods differ significantly. This difference occurs because of the solidifying particles. While the enthalpy method calculates a continuously increasing particle temperature for increasing particle sizes, by the separation method a local minimum is calculated for particle sizes between 50 and 60 μm in diameter. In addition, the solid fraction shows a greater gradient when calculated by the separation method; here the particle size spectrum of the solidifying particles is smaller in the separation method. With respect to the thermal state of the particles, this means that all particles in the size spectrum 50–110 μm are, on average, slightly undercooled. This can be seen from the mean particle temperature, which is lower than the liquidus temperature (upper limit of the solidification interval – grey area) though the mean solid fraction is still zero. The reason for the different results of both averaging methods in this particle size spectrum is explained as follows. If a fluid melt droplet is undercooled, its specific enthalpy is lower than the specific enthalpy at liquidus temperature. Because of this, when averaging with the enthalpy method, this droplet will be taken as a solidifying particle, while in fact this droplet is still fully liquid. In the separation approach, the solid fraction is calculated separately from temperature and, therefore, the mean solid fraction is derived from the particle's fluid melt content only.

In contrast to the properties exhibited by thermal average of individual particle-size classes at a certain position within the spray, the averaging methods above yield strongly deviating results when locally averaged over the whole particle-size spectrum in the spray at a certain (radial) position. This behaviour is shown in Figure 5.34. When averaging over a certain particle-size class (bin), the mass fraction of each individual particle parcel is identical (as derived from the spray model). But when averaging over the whole particle-size spectrum, the mass distribution of the particle-size distribution is to be considered. When looking at a particle mass where half of the mass consists of small, cold and fully solidified particles, and the other half consists of a single big, fully fluid and hot particle, then the separation method yields a mean solidification state as a solid fraction of $f_{s,m} = 0.5$ (independent of individual temperatures or the latent heat content of the single hot particle). For the enthalpy method, the mean properties are calculated in a coupled way. Here, the resulting mean temperature and the mean solid fraction depend on the difference between the temperature of the small particles and the solidus temperature compared to the remaining superheat within large particles.

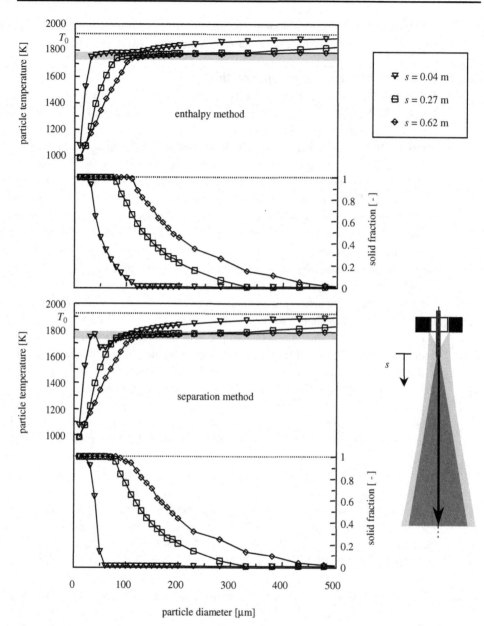

Fig. 5.33 Comparison between two averaging methods for particle temperatures and solid fractions, mean averages for different particle diameters and flight distances *s*. Calculation parameters are those noted in Table 5.5 (Bergmann, 2000)

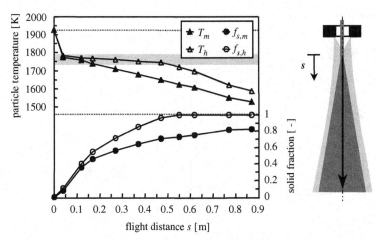

Fig. 5.34 Mean temperature and solid fraction of the particle mass along the flight path (index: m = separation method; h = enthalpy method). Calculation parameters used are shown in Table 5.5 (Bergmann, 2000)

The differences between the two averaging methods are visible in Figure 5.34, and can be interpreted as follows. The enthalpy method represents the thermal equilibrium with respect to the particle mass in the spray. This is illustrated by bringing together all particles from a specific location (point of impingement) and leaving this mass under adiabatic conditions for inner compensation processes (heat conduction and solidification) to obtain the equilibrium values of T_h and $f_{s,h}$. In contrast, the separation method describes the instantaneous local thermal state (T_m and $f_{s,m}$) of the particle mass in the spray. As can be seen from Figure 5.34, the instantaneous mean particle temperatures and mean solidification fractions are below the values for the equilibrium state (enthalpy method) for the whole particle flight distance. With respect to the solidification process, after the deposition this difference means that the overall particle mass is undercooled and will be reheated (in the deposit) by means of the latent heat released during solidification.

From comparison of the two different thermal averaging methods with respect to the thermal properties of the particle mass located at the spray centre-line, the spray may be divided into three different zones:

- Zone 1, overheated spray ($T_m > T_h; f_{s,m} > f_{s,h}$). In this zone, the values from the separation method are greater than those from the enthalpy method. For the overall particle mass, this means that the particle material tends towards lower solidification fractions at the thermal equilibrium. Thus, if the particle mass is brought together as is done during deposition, the remaining superheat from the bigger, still liquid droplets will remelt parts of the already solidified particle mass.
- Zone 2, slightly undercooled spray ($T_m < T_h; f_{s,m} < f_{s,h}$). In this zone the values from the separation method are lower than those from the enthalpy method. The averaged particle temperature (with respect to specific enthalpy) is in the range of the solidification interval ($T_s < T_h < T_l$). Therefore, the particle material tends towards higher solid fractions during

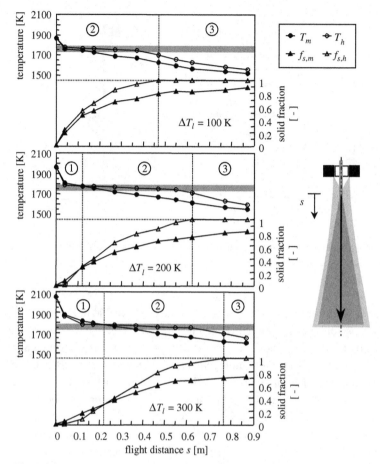

Fig. 5.35 Comparison of different averaging methods for varying melt superheats, showing mean temperature and solid fraction of the whole particle mass along the flight path at the spray centre-line (dependent on the melt overheat). For other parameters see Table 5.5 (Bergmann, 2000)

thermal equilibrium. The remaining superheat in the bigger liquid drops is not sufficient to remelt parts of the already solidified particle material. The particle mass still contains enough enthalpy that even during thermal equilibration some melted parts will remain ($f_{s,h} < 1$).

- Zone 3, heavily undercooled spray ($T_h < T_s$; $f_{s,h} = 1$). In this zone the particle material at thermal equilibrium is completely solidified ($f_{s,h} = 1$). The equilibrium temperature is lower than the solidus temperature ($T_h < T_s$). In this zone, the deposited material will be immediately solidified.

Based on adjustment of the melt superheating in the crucible/tundish, the boundaries between these three zones can be shifted relative to the flight distance of the particles, as shown in Figure 5.35. For a small amount of melt superheating ($\Delta T_l = 100$ K), cooling of the particles occurs immediately after solidification. For this case, Zone 1 is not visible. The transition from Zone 2 to Zone 3 is approximately at a flight distance of $s = 0.47$ m.

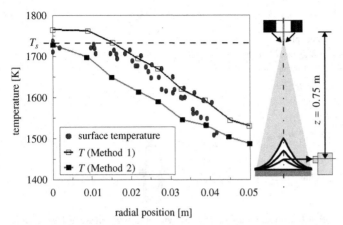

Fig. 5.36 Radial distribution of the averaged temperature at the deposit surface (Bergmann *et al.*, 1999, 2000): Method 1, equilibrium; Method 2, separation. Measurements taken using a pyrometer (Kramer, 1997)

By increasing the melt superheating, the transition is shifted to increase flight distances ($s = 0.62$ m for $\Delta T_l = 200$ K and $s = 0.77$ m for $\Delta T_l = 300$ K). In these cases of increased superheating, Zone 1 is established.

A comparison of results obtained by these two different averaging methods with respect to the surface temperature of a spray formed steel (C3) deposit is shown in Figure 5.36. In the same figure, a temperature distribution measured by means of a surface pyrometer (Kramer, 1997) is shown. Both methods obtain a monotone, and at higher values of the radial coordinate attain almost linear behaviour of the surface temperature with increasing radius. The values of the surface temperature calculated by the equilibrium method are greater than those calculated by the separation method. The measured values of surface temperatures are located in between both calculated curves.

Coupling effects between phases have to be taken into account during spray simulation within thermal spray processes, as can be seen from another result by Bergmann *et al.* (2000). In Figure 5.37, once again, the calculated mass flux distribution in the spray at the point of impingement onto the substrate for three different boundary conditions is illustrated:

(a) without coupling effects;
(b) with coupling, taking the standard Ranz–Marshall (1952) correlation of Nusselt numbers for heat transfer calculation of the droplets (see Eq. (5.16));
(c) with coupling, taking the extended Nusselt number correlation from Yearling and Gould (1995), including the turbulent character of the gas flow field (see Eq. (5.18)) for the heat transfer calculation.

From the shown distribution of mass one can clearly see that coupling effects directly influence the macroscopic spray behaviour in terms of the spray angle. Without coupling, the spray angle is just half of the value with coupling, and the simulated spray will be distributed in an unrealistically narrow spectrum.

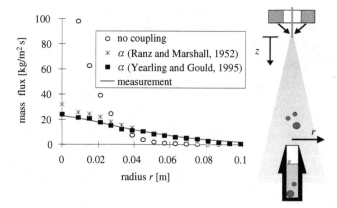

Fig. 5.37 Mass flux distribution in spray, calculated with and without regard for coupling effects (Bergmann *et al.*, 2000)

5.3   Spray-chamber flow

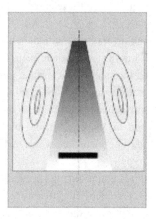

Within all spray processes that aim to impact a droplet mass onto a substrate, the inner spray flow field within the main spray cannot be analysed properly without recourse to the geometry of the outer facility that contains the spray and the type of external flow, as well as the flow field in the surrounding environment. In spray forming applications, this area of study involves the analysis of the flow conditions in the spray chamber. The influence of external flow conditions on the primary spreading behaviour of the spray plume is not relevant in most cases (at least when some principal fluid mechanical rules are recognized correctly). But investigation and optimization of the spray-chamber geometry from a fluid mechanical point of view allow control of some important process conditions and features. In addition, at least one degree of freedom is added for influencing the spray cone behaviour in the process: in spray forming this is, for example, control of the overspray particulate mass, which is transported in the spray chamber and possibly reentrained into the spray. Additional possibilities of process control and manipulation are based on measures of active

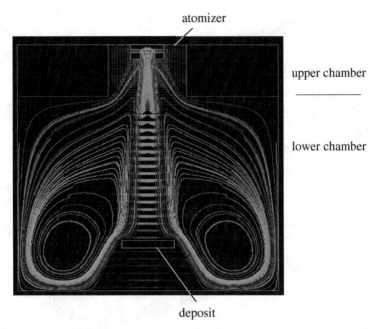

atomizer

upper chamber

lower chamber

deposit

Fig. 5.38   Spray-chamber flow: stream-line representation of the gas flow field in the lower and upper chamber

flow steering in the spray chamber. Here active elements, such as secondary gas flows (e.g. by some gas jets in the spray chamber not participating in the atomization process, but just controlling the chamber flow) in the chamber may be applied (without influencing the principal atomization and spray development process). Or passive control may be achieved, e.g. by applying flow baffles in the chamber. This flow control feature may be of importance for spray forming applications as recirculating and swirling gas flows in the spray chamber contain a certain amount of overspray powder. Here process problems may occur due to:

- radial entrainment of solidified powder into the spray (which on the other hand may even be, in some cases, advantageous for additional cooling of the spray);
- impairment of optical spray control and measurement devices for process monitoring or control;
- possible sticking of overspray particles in specific areas of the spray chamber that afford cost-intensive and time-consumptive cleaning processes (and may cause hazardous risks due to increased explosion potential).

Potential for the development of spray forming process by passive and active control of flow in the spray chamber may be based on modelling and numerical simulation of the two-phase flow; this potential has not been generally realized yet. The possible reason for this is the huge computer resources needed to calculate the fully three-dimensional flow field in the chamber. The two examples listed next illustrate the status quo in process developments.

The simulated flow field in a cylindrical spray forming chamber, based on a two-dimensional coupled two-phase flow simulation (Fritsching, 1995), is illustrated in Figure 5.38 in terms of gas stream lines and gas velocity vectors at certain positions. The

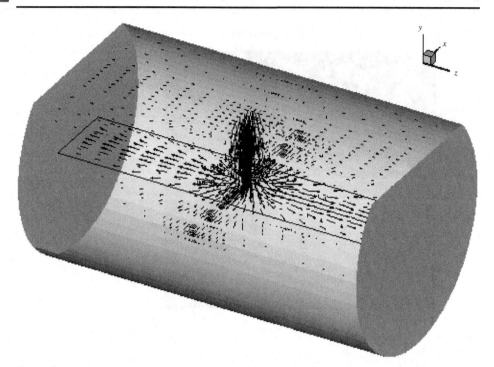

Fig. 5.39   Spray-chamber flow during spray forming of aluminium: gas flow field (Pien *et al.*, 1996)

chamber under investigation consists of two cylindrical segments. The geometry of the real spray chamber that this has been aimed to reflect was rectangular, but has been simplified to that of a cylinder for simplicity of calculation, as the simulation can then be performed in two dimensions. The atomizer is located in the upper segment, at the top of the spray chamber. One can observe marked recirculatory flows of gas, which cause entrainment of gas and continuous recirculation of finer overspray powder into the spray within the upper part of the system. This reevaluation of flow from the lower spray chamber back to the upper atomizer chamber needs to be avoided. In a later version of this particular spray chamber, the atomizer is mounted at a lower position directly inside the main spray chamber to avoid such problems.

Figure 5.39 shows the simulated gas flow field in a spray chamber from a study by Pien *et al.* (1997). The coupled two-phase flow simulation has been adapted for a three-dimensional calculation in the spray chamber. The process under investigation aims at the production of flat products by spray forming. Therefore, the base chamber geometry has been constructed within a relatively long circular chamber. Figure 5.40 illustrates an instantaneous picture of the drop distribution in the spray chamber for the same simulation run, where the actual positions of the calculated parcels from the simulation are to be seen. The investigation has been made during spray forming of aluminium alloys where the melt is atomized by flat slit (linear) nozzle systems. Here the melt exits the tundish in the form of an elongated flat sheet and is atomized by means of the so-called linear atomizer attacking the liquid sheet from both sides. The aim of this investigation is to visualize the flow field in

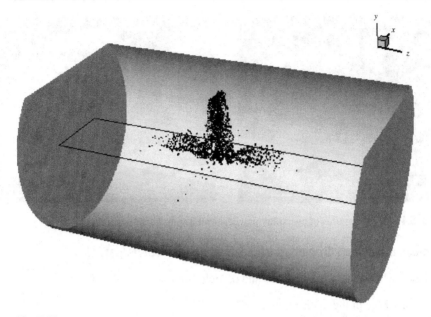

Fig. 5.40 Spray chamber flow during spray forming of aluminium: particle movement (Pien *et al.*, 1996)

the chamber and to optimize the positioning of the nozzle and exhaust gas system within the spray chamber. The figures show one calculated alternative where the exhaust system is located on both sides of the spray chamber at half height.

5.4   Droplet and particle collisions

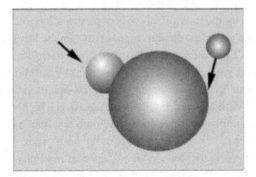

Collisions between particles and solid boundaries (wall impact) or between particles and particles (collision) have a great impact on the behaviour of dispersed multiphase flows such as sprays. These collision processes must be incorporated as a submodel in a complete analysis of spray modelling.

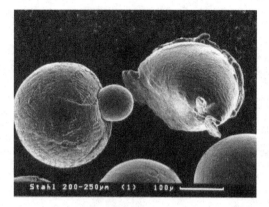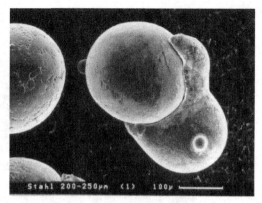

Fig. 5.41   SEM pictures of overspray powder from a spray forming facility

5.4.1   Collision between droplets/particles and other droplets/particles

Binary collision between droplets and particles in a spray may lead to:

- coalescence of the collision partners;
- disintegration of one or both collision partners;
- adhesion or sintering of particles to clusters, for particles which have been solidified or partly solidified before collision.

In all cases, the local spray structure is changed due to collision, by means of the local drop-size distribution, which will change, and also the local drop-vector momentum distribution, which will change. Compared to conventional spray processes, in the special case of molten metal sprays and spray forming applications, additional morphologies of spray droplets and particles have to be included in collision modelling. Here, besides fluid droplets, partly solidified particles and solid particles may act as collision partners. These pairings may lead to a variety of different phenomena during particle/droplet collision in molten metal droplet sprays. The SEM photographs in Figure 5.41 show a sample of an overspray powder taken from a steel spray forming process: the fraction shown is of particles 200 to 250 µm in size, and illustrates binary droplet and particle collisions in the spray. On the left-hand side of the figure, a small solid particle is illustrated, which has partly penetrated into another particle during impact. This collision partner has obviously been partly solidified at the moment of collision. Another collision event is observed in the right-hand figure. Here a partly solidified droplet has collided with another droplet of similar size. The first droplet was in a partly solidified state during collision and contained a liquid core that has poured over the second particle during impact.

The relevance of droplet collision processes to a model of dispersed multiphase flow is dependent on several process parameters. The greatest influence from these collision processes is expected in highly concentrated flows and within flow fields with large differences in the velocity spectrum of the dispersed phase. For example, in highly dispersed flows, the velocity spectrum that needs to be evaluated for droplet collision events (assuming that the particle density is not too high) is mainly affected by the turbulent fluctuations of the particle velocities. Here collision effects are most likely. Within spray forming, turbulent effects on the movement of particles are of minor importance as the metal particles normally have high material densities and, therefore, high inertial effects will govern particle motion (excluding, perhaps, aluminium sprays). Also, the high gas velocity gradient in the spray will contribute to this effect. In spray forming, the resulting velocity spectrum within the droplet-size distribution between smaller and bigger particles and their respective slip velocities to the gas phase mainly determine the relevance of droplet collision events.

A rough estimate of the relevance of particle collision effects to the overall spray behaviour can be obtained from evaluation of the specific ratio of particle relaxation time (acceleration dominant) τ_p and the time scale between individual particle collisions τ_k (Crowe, 1980). The particle relaxation time is a measure of the time that a particle takes to accelerate from velocity v_1 to v_2. The particle relaxation time may be simply expressed by means of the Stokes number:

$$\tau_p = \frac{\rho_p d_p^2}{18\mu_f}. \tag{5.61}$$

The time constant between collisions for particles is derived from the collision frequency:

$$\tau_k = \frac{1}{f_k}. \tag{5.62}$$

In dispersed multiphase flows at low particle concentrations (dilute flow), particle transport is mainly determined by fluid dynamic interaction of individual particles with the continuous carrier phase (e.g. drag, lift, etc.). At high particle concentrations (dense flow), particle collision mainly affects the movement of individual particles. These two regions are separated by a characteristic time scale:

$$\text{dilute: } \frac{\tau_p}{\tau_k} < 1, \quad \text{dense: } \frac{\tau_p}{\tau_k} > 1. \tag{5.63}$$

In a dense dispersed two-phase flow, the time distance between particle collisions is smaller than the particle relaxation time. Before reaching another steady slip velocity, from droplet gas interaction, another collision will occur. Therefore, particle movement is determined mainly by collision, and fluid dynamic effects are of less importance. Obviously, in dilute two-phase flows particle collisions will also occur, but their probability is small and the main flow and particle behaviour is not strongly influenced by collisional effects.

A suitable binary collision model in a dispersed two-phase flow must be able to describe the following events and properties:

(1) When (or, if a correlation between place and time exists, where) does a particle collision occur?
(2) Which particles (collision partners) are employed in the collision process?
(3) What happens during collision to the collision partners?
(4) What is the outcome of the collision?

The most plausible solution of collision modelling within a two-phase flow simulation is based on direct and simultaneous tracking of all particles in the flow field in terms of their individual movements and momentary positions. Here the occurrence of a collision and the collision partners themselves may be directly identified from simple geometric relations, as the position and velocity vectors of all particles at any time are known in the computation. But the computational effort for this approach is very high. This manner of collision modelling is, up to now, only possible in quite simplified flow configurations (see O'Rourke (1981)).

Computationally easy models for calculation of binary collisions within multiphase flow are based on an analogy of collision processes to processes in kinetic gas theory. Therefore, collision probabilities will be derived which indicate the occurrence of a binary collision. In a stationary flow field, the collision probability of a particle in a finite volume is calculated from the mean droplet concentration and relative velocity in that volume. The number of collisions of Particle 1 with any other, Particle 2, in a control volume of size ΔV within a time interval Δt is given by:

$$P = f_k \Delta t = \frac{\pi (d_{p1} + d_{p2})^2 |\mathbf{u}_{p1} - \mathbf{u}_{p2}| n_p \Delta t}{4 \Delta V}, \tag{5.64}$$

where n_p is the number of particles in the control volume and the collision projected area is $\pi/4(d_{p1} + d_{p2})^2$ and includes the value of the relative velocity between particles.

A first estimate of the collision probability between particles in a spray during spray forming applications is done from a coupled simulation run. Here, simulation for an atomizer gas pressure of $p_2 = 2.5$ bar (described in Chapter 5.2) is used. This calculation assumes a droplet-size distribution which fits the log-normal type with a mass median value of particle sizes $d_{50.3} = 80$ μm and a logarithmic standard deviation of $\sigma_{\ln} = 0.7$. From the calculated particle trajectories in the spray, at first the local number, or concentration value, of particles in each control volume is calculated from:

$$N = \sum_i \frac{6\dot{m}_{p,i}}{\pi d_{p,i}^3 \rho_p u_{p,i}}. \tag{5.65}$$

Here all local particle size classes i are summed up. The first result of the analysis is illustrated in Figure 5.42. The vertical axis is the nozzle distance; it starts below the fragmentation area at $z = 0.08$ m. The maximum concentration number in that spray area close to the centre-line in the fragmentation area is $1.4 \times 10^{10}/\text{m}^3$ particles. For calculation of the collision frequency of a specific particle within the spray, a model is assumed for the whole spray where:

(a) the representative mean droplet size in the spray is the median of the number density distribution of the particle-size distribution $d_{50.0} = 17$ μm, which is calculated from the mass median diameter by $d_{50.0} = d_{50.3} \, e^{-3\sigma^2}$ for the assumed log-normal drop-size distribution;

(b) for the representative relative velocity, a local-averaged maximum relative velocity (averaged for all particle-size classes) has been assumed.

From this model, the value of the maximum possible collision frequency of the specific particle under investigation in the spray is:

$$f_{\max} = \frac{\pi}{4}(2d_{50.0})^2 u_{\text{rel}/\max} N. \tag{5.66}$$

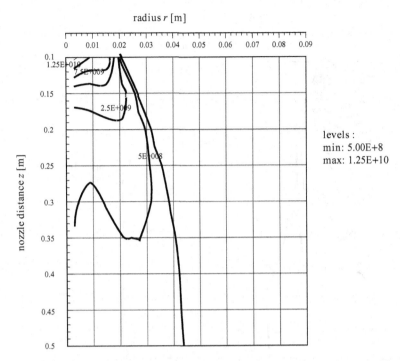

Fig. 5.42 Number concentration of metal melt particles in the spray: $p_2 = 3.5$ bar, material, steel C35; metal mass flow rate $= 0.192$ kg/s; melt super heating $= 120$ K; $d_{50.0} = 17$ μm

The calculated maximum collision frequencies are illustrated in Figure 5.43 as isolines. Maximum collision frequencies in the fragmentation area are of the order of 3000/m³.

Based on collision probability, Eq. (5.62), at each time step of the numerical particle tracking algorithm, comparison with a stochastic random process decides whether or not a collision will take place (see Sommerfeld (1995)). As a collision partner for the binary collision model of the specific particle under investigation, a fictitious partner particle is sampled from the local particle-size spectrum, with properties randomly determined from an assessment of all particles in that particular cell volume.

In the last step of the collision model, the result or outcome of the binary collision event needs to be predicted. If both collision partners are solid particles, the outcome of the collision may be easily described by means of elastic or plastic collision, based on momentum equations. For fluid droplets, the physics is more complicated as coalescence or disruption of the droplets may occur. Georjon and Reitz (1999), for example, analysed the formation of a common cylindrical liquid element as a result of binary collision and coalescence of two droplets at a sufficiently high Weber number. The liquid cylinder may disintegrate, due to capillary-induced instabilities, into droplets. The main parameters of a binary droplet collision model are (see Figure 5.44 for relevant values): the relative velocity, the size and properties of both droplets (which are combined within the local Reynolds and Weber numbers) and the impact parameter χ. The impact parameter describes the shortest

radius r [m]

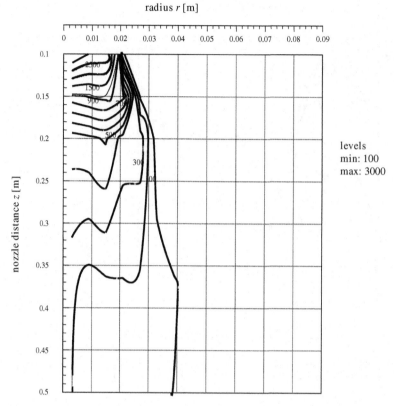

nozzle distance z [m]

levels
min: 100
max: 3000

Fig. 5.43 Maximum collision frequency of the mean particle size (parameters as in Figure 5.42)

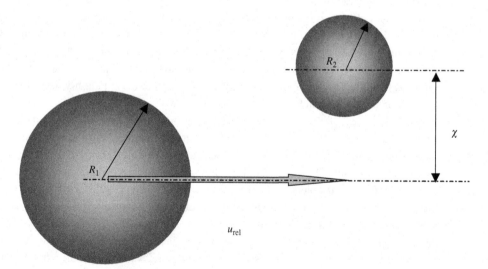

Fig. 5.44 Binary in-flight collision of droplets: relative velocity u_{rel}, impact parameter χ

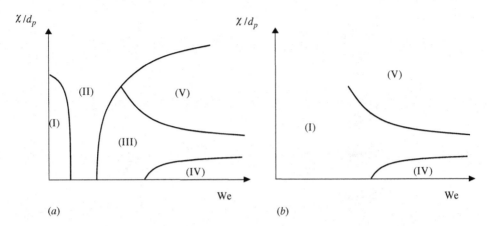

Fig. 5.45 Event chart of droplet collisions for (*a*) fuel, and (*b*) water at ambient pressure (Qian and Law, 1997)

distance between the tangential trajectories of two equally sized droplets at the moment of collision.

A stochastic model for incorporation of particle/particle collisions during simulation of a dense spray based on Lagrangian tracking of particles has been introduced by O'Rourke (1981).

The basic phenomena of coalescence and separation of droplets during binary collision have been experimentally tackled by Ashgriz and Poo (1990) and Qian and Law (1997) for water and fuel droplets, and Menchaca-Rocha *et al.* (1997) for mercury droplets.

Qian and Law's (1997) investigations have led to a collision event chart for water and fuel droplets that is illustrated qualitatively in Figure 5.45. The different boundaries between the collision modes are described based on the impact Weber number and the impact parameter χ. Qian and Law recognize five distinct regions or collision modes:

- In Area I, the colliding droplets coalesce immediately after only small deformation.
- At somewhat increased Weber numbers, in Area II the droplets separate after collision and are repelled from each other, as the time for drainage of the gas film in between colliding droplets is insufficient.
- In Area III, the increased relative collision velocity between the droplets drains the gas film out of the gap between the droplets, and the droplets coalesce once again. In this area, the coalescing droplets are strongly deformed during collision.
- At lower impact parameters in Area IV, where most central collisions occur, the droplets first coalesce and then separate in a subsequent collision stage.
- For a sliding or non-central collision, in Area V, the collision behaviour is similar to that in Area IV, but because of the sliding impact the probability of satellite droplet production by partial fragmentation is higher.

These regimes have been found for fuel droplets: for water droplets in the investigated range of impact parameters, only modes I, IV and V have been found.

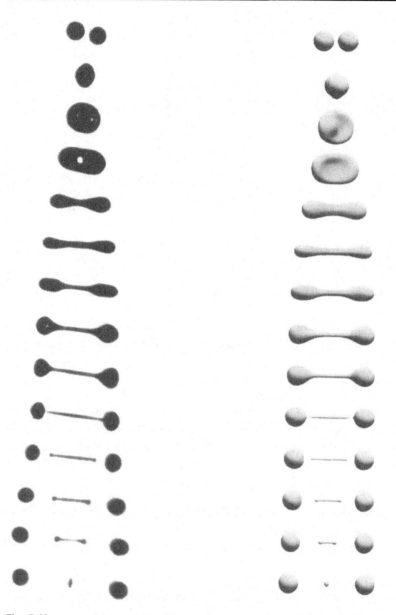

Fig. 5.46 Comparison of lattice Boltzmann simulation result with experiment for binary liquid droplet collision: $\chi = 0.5$, Re $= 100$, We $= 106$ (Frohn and Roth, 2000)

Direct numerical simulation of the behaviour of two droplets during collision is performed, for example, in the work of Nobari and coworkers (1996a,b) and Frohn and coworkers (see, for example, Frohn and Roth (2000)). Some principal physical details can be explained by this numerical approach. Figure 5.46 shows a comparison between a lattice Boltzmann simulation and experiment for binary collision of droplets (Frohn and Roth,

2000). The parameters used for the simulation are: We $= 106$, impact parameter $\chi = 0.5$ and Re $= 100$. The colliding droplets have the same size. The results of the numerical simulation, especially the formation of satellite droplets, are in excellent agreement with experiment.

In his description of binary collision between droplets, O'Rourke (1981) distinguished between gracing or sliding collisions and a collision proper, which finally leads to permanent coalescence between participating droplets. The criterion for the difference is the critical collision angle Φ_{crit}, which is given for water droplet collisions as:

$$\sin^2 \Phi_{\text{crit}} = \min \left[2.4 \frac{f(d_{p,1}/d_{p,2})}{\text{We}} ; 1 \right]. \tag{5.67}$$

Here the correlation $f(d_{p,1}/d_{p,2})$ has been derived and adapted from the experimental investigations of Amsden et $al.$ (1989). A polynomial correlation is used and is dependent on the diameter of the droplets participating in the collision. If the actual collision angle is smaller than the critical collision angle Φ_{crit}, the droplets show a central collision and will coalesce, otherwise the collision is a sliding one.

The velocity of a droplet after a sliding collision can be deduced from:

$$u_1' = \frac{u_1 m_1 + u_2 m_2 + m_2 (u_1 - u_2)[(\sin \Phi - \sin \Phi_{\text{crit}})/(1 - \sin \Phi_{\text{crit}})]}{m_1 + m_2}, \tag{5.68}$$

where the direction of movement of the droplet remains unchanged from the direction of the droplets before collision. The velocity of the combined droplet after coalescing droplet collision is:

$$u_1' = \frac{u_{1,i} m_1 + u_{2,i} m_2}{m_1 + m_2}; \quad i = 1, 2, 3. \tag{5.69}$$

Here the diameter of the resulting droplet is calculated from addition of the volumes of the individual collision partners:

$$d_{p,\text{new}} = \left(d_{p,1}^3 + d_{p,2}^3 \right)^{1/3}. \tag{5.70}$$

Application of a collision model during simulation of a propane spray is documented by Aamir and Watkins (1999). Application of this collision model for analysis of a spray from a pressure swirl nozzle has been performed by Rüger et $al.$ (2000), who compares the simulation results to detailed measurements. In the area of the dilute spray flow, only an insignificant influence of collision events on the distribution of the mean number diameter $d_{50.0}$ in the spray has been detected. But a remarkable effect on the integral Sauter diameter $d_{3.2}$ in the direction of the developed spray has been found. By taking into account coalescence events by the collision model, the integral Sauter diameter is increased and the integral number density flux within the spray is decreased. These numerical results are in good agreement with experimental findings.

Application of a binary collision model within spray simulation in spray forming applications or powder metal production by melt atomization is an actual research task. In the latter, for example, the formation of satellite droplets (smaller particles sticking at the surface of bigger particles as a result of drop collisions) is a common problem in powder production as it increases the amount of non-spherical particles in the product. This particle property may cause some problems in subsequent operation processes with the atomized powder (reduced flowability, etc.). But in the area of droplet collision with varying droplet morphologies (i.e. solid, semi-solid, liquid), information is not sufficiently available to derive a general collision model within melt atomization and melt spray simulation.

5.4.2   Collision of melt droplets with solid particles (ceramic)

An important advantage of the spray forming process is the possibility of producing metal-matrix-composites (MMCs). For example, by adding ceramic particles (such as SiC, Al_2O_3, TiB_2 or TiC) or non-metallic powders (e.g. graphite) directly in the spray, a homogeneous combination of a metal matrix plus composite inserts in the deposit can be produced. These spray formed MMC materials exhibit no segregation, as seen in conventionally produced MMC material, and have improved material properties when compared with the base material of the matrix. Thereby volume concentrations of up to 20% solids may be introduced into spray formed preforms via particles in the spray. These particles may be conveyed into the spray within the atomization gas or via a separate gas-assisted delivery system. For analysis and derivation of the expected local distribution of solid contents in the matrix, it is necessary to investigate the way in which the solid particles are incorporated. Here the interaction mechanisms during particle/droplet collision need to be described.

Experimental investigations indicate that different interaction mechanisms effectively influence the incorporation of ceramic particles into the matrix material. Three possible mechanisms are illustrated in Figure 5.47 based on the work of Gupta *et al.* (1991). For low relative velocities between the liquid droplet and the solid particle, the kinetic energy for penetration, needed to overcome the surface tension of the droplet, is not sufficient. Impact may lead to adhesion of the solid particle onto the drop surface. The adhered particles may be completely incorporated into the matrix upon impact of the liquid droplet onto the deposit if the impact occurs at a sufficient velocity (see Figure 5.47(a)). In the second mechanism, solid particles are deposited directly on the surface of the preform. These may be incorporated in the matrix if liquid droplets subsequently impinge at that specific location (see Figure 5.47(b)). In Figure 5.47(c) the solid particle penetrates into the liquid droplet after collision in flight; this is the third particle inclusion mechanism. Here a minimum kinetic impact energy for particle penetration is needed.

Physical models for description of the particle penetration mechanism during injection of solid particles into the spray are based on energy and/or force balances (see, for example, Majagi *et al.* (1992), Wu *et al.* (1994) and Zhang (1994)). These models have been summarized by Lavernia and Wu (1996). Basic assumptions and preconditions for a collision model are that:

- the solid particles are smaller in size than the liquid droplets, by at least one order of magnitude;
- within the solidification process of the molten droplet, the morphology (e.g. the viscosity) of the drop material changes and therefore the penetration potential of the particle changes.

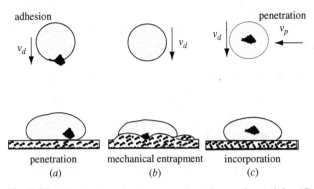

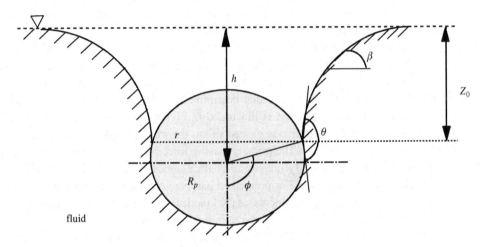

Fig. 5.47   Mechanisms for incorporation of ceramic particles (Gupta *et al.*, 1991)

Fig. 5.48   Penetration model (Majagi *et al.*, 1992)

Majagi *et al.* (1992) derived a force balance for the impact and penetration of a solid particle on a two-dimensional liquid surface in the case of a fully liquid melt droplet. The droplet is moving in the direction of gravity. This model is illustrated in Figure 5.48. R_p is the radius of the solid particle, r is the radius of the meniscus, h is the penetration depth of the centre of gravity of the solid particle below the liquid surface of the drop and Z_0 is the penetration depth of the contact line. The angles β, ϕ and φ are the angles of the meniscus, the wetting angle and the angle towards the contact point, respectively. The model is based on the assumption that the resistance against penetration of the particle is due to surface tension and hydrostatic pressure. The force from the surface tension using the above terms is formulated as:

$$F_\sigma = -\pi d_p \sigma_l \sin \phi \sin(\phi + \theta). \tag{5.71}$$

The hydrodynamic pressure due to buoyancy in the submerged part of the particle minus the part of the hydrostatic pressure in the gas environment above the penetrating particle is:

$$F_h = -\frac{\pi}{24}d_p^3\rho_l g(1 - \cos\phi)^2(2 + \cos\phi) + \frac{\pi}{4}d_p^2 g\sin^2\phi\,\rho_p Z_0, \tag{5.72}$$

and the force due to gravity is:

$$F_g = \frac{\pi}{6}d_p^3\rho_p g. \tag{5.73}$$

In this model, the change in velocity of the particle during impact is associated with a force F_v:

$$F_v = \frac{\pi}{12 P_d}d_p^3\rho_p\left(v_0^2 - v_1^2\right), \tag{5.74}$$

where the velocities of the particle before (v_0) and after (v_1) impact are used. It is assumed that when the force balance results in a remaining force in the direction of gravity, complete penetration of the particle into the fluid surface occurs.

This model has been used by Majagi *et al.* (1992) to describe the penetration behaviour of ceramic particles in fluid metal surfaces during melt atomization. These authors derived as the main influencing parameters: the density of the particle, the contact angle (wetting angle) and the solid particle size that determine the penetration process. An example of the results of this investigation is illustrated in Figure 5.49. Here the minimum relative velocity needed for penetration is plotted versus the wetting angle (a), the particle density (b) and the particle size (c). The primary driving force for penetration is the kinetic energy of the impinging particle. By increasing contact angles, the necessary minimum velocity for penetration is increased. For prescribed particle velocities, a higher particle density or an increased particle diameter increases the kinetic impact energy. Therefore, the minimum necessary penetration velocity is lowered in this case.

The model of Majagi *et al.* (1992) is limited to the analysis of penetration processes within fully liquid surfaces. In the area of melt atomization and spray forming, the change in particle material properties (morphology) related to solidification of the droplet in the spray from liquid via semi-solid to solid is of specific interest. As has been demonstrated from the results in Chapter 5.2, the main part of the particle mass in the spray, in spray forming, is in the state of phase change. A model describing the morphology of droplets dependent on the solidification status has been derived by Wu *et al.* (1994) and Zhang *et al.* (1994). These authors introduce, in addition to the penetration resistance due to surface tension, the fluid dynamic resistance of the particle during penetration into the liquid drop and the movement of the liquid matrix. The latter is described by the resistance force as:

$$F_w = \frac{1}{2}\rho_l u_p^2 A_p c_d \tag{5.75}$$

for different solid particle geometries such as a sphere, square and rhomboid, and with the correlation of the resistance coefficient dependent on the Reynolds number, from Eq. (5.4). For evaluation of the model these authors chose the approximate relations during impact of Al_2O_3, graphite, SiC and TiB_2 particles in a fully liquid or semi-solid aluminium

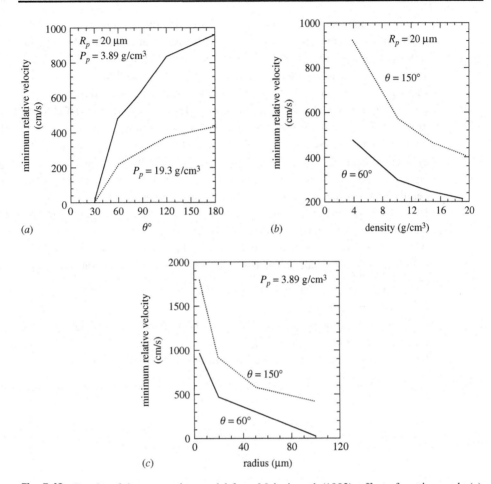

Fig. 5.49 Results of the penetration model from Majagi *et al.* (1992): effect of wetting angle (*a*), particle density (*b*) and particle size (*c*)

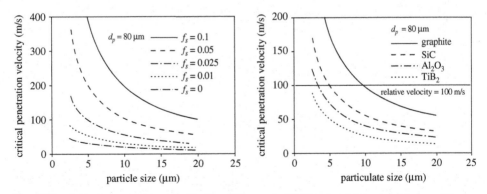

Fig. 5.50 Critical penetration velocity from Wu *et al.* (1994), aluminium droplet $d_p = 80$ μm, solidification ratio and SiC particles (left), different solid particles (right)

droplet as the boundary condition. The results of this model indicate a small variation of the penetration potential with respect to particle form (at constant total volume) for the spherical, quadratic and rhomboid-shaped ceramic particles.

The critical penetration velocity for different boundary conditions from Wu *et al.* (1994) is summarized in Figure 5.50 for an aluminium droplet having a diameter $d_p = 80$ μm. On the left-hand side of the figure, the results for different droplet solidification ratios f_s between 0 and 10% are illustrated for the impact of a SiC particle. As expected, the critical penetration velocity is drastically increased with increasing solidification ratio even at these small degrees of solidification of the droplet. On the right-hand side of the figure, comparison of the critical penetration velocities for different penetrating solid particle materials is shown. The solidification rate is assumed to be $f_s = 0.025$ in this figure. In the order of decreasing solid particle densities and increasing contact angles, the penetration ability increases from graphite via SiC, Al_2O_3 to TiB_2.

In summary, for the penetration ability of solid particles (e.g. ceramics) during impact onto melt droplets in the area of melt atomization or spray forming one can state:

- for an unsolidified liquid droplet, the contact angle and the particle density are most important for penetration;
- for a semi-solid droplet, during solidification the flow resistance inside the droplet is important and, therefore, the solidification ratio of the droplet and the solid particle density are the main determining properties of the penetration process.

6 Compaction

Description of the compacting process and the resultant behaviour of the spray deposited layer, or spray form, are key parameters for all impact-orientated spray processes, as this is the stage during which all remaining product properties are determined.

For modelling and simulation of the transient growth, the solidification process and the temperature distribution inside the deposit/substrate, the main boundary conditions influencing this thermal process are prescribed by the process parameters. These boundary conditions are:

- development of a local impacting droplet mass and the geometric construction of the deposit;
- analysis of the droplet impact processes, contributing to the deposited mass or the over-spray, or the establishment of porosities in the deposited layer;
- the thermal state of the spray during impact (in terms of solidification ratio and particle enthalpy flux);
- the heat flux from and to the surrounding atmosphere from the deposit and substrate surface, as well as the heat flux from the deposit to the substrate.

Most of the major boundary conditions and submodels of this thermal stage are derived from the numerical simulations presented earlier.

6.1 Droplet impact and compaction

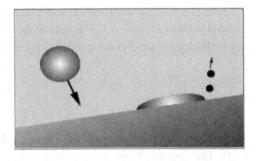

The behaviour of impinging droplets is a key parameter in spray and droplet-based manufacturing techniques. Here, models for drop impact behaviour, not only in spray forming but

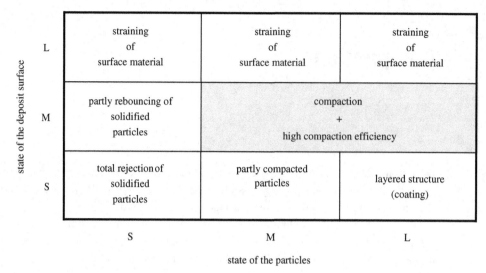

Fig. 6.1 Probability of impact results with spray forming (Mathur *et al.* 1989b): L, liquid; M, mushy – partly solidified; S, solid

also for application within single-drop processes, e.g. for rapid prototyping, are of importance. A review of thermo-fluid mechanisms controlling droplet-based materials processes has been given by Armster *et al.* (2002).

Only part of the total mass impacting on the substrate or deposit from the spray compacts and contributes to the growth of the deposit during spray forming. Dependent on substantial substrate (or deposit, after initial coating of the substrate) process conditions (kinetic and thermal) and the impinging droplets in the spray, these impinging particles may:

• be completely deposited;
• be totally reflected from the surface;
• extract fragments of the already deposited material, which may be partially leached;
• be fragmented during impact and partly reflected.

The influence of the thermal state of the particle substrate on compaction behaviour has been studied by Mathur *et al.* (1991). A qualitative model has been derived which divides the compaction into different regimes, and is shown in Figure 6.1. The state of the droplets during impact and the state of the surface of the deposit are divided into different categories:

• liquid or fully liquid (L);
• semi-solid or mushy (M);
• and solid, fully solidified (S).

Dependent on the pairing of thermal properties from the droplets and the surface, different compaction conditions may be observed. The best condition for maximum compaction efficiency is a semi-solid deposit surface impinged by fluid or partly solidified particles.

For a detailed analysis, in addition to the thermal state identified in Figure 6.1, the kinetic state of the particles in terms of velocity, impact angle and particle size also needs to be accounted for.

The amount of non-compacting or reflected material contributes to the overspray of the process, which can be quantitatively described by the local ratio between the compacting mass fluxes $\dot{m}_k$ (contributing to the product) and the impinging metal mass flux from the spray. This yields the compaction rate k_p:

$$k_p = \frac{\dot{m}_k}{\dot{m}_s}.$$
(6.1)

The locally and temporally changing value of the compaction rate during spray forming is influenced by a number of parameters. Those parameters that best influence the formation of Gaussian-shaped deposits by spray forming have been investigated experimentally by Kramer *et al.* (1997) and, more generally, by Buchholz (2002). From these investigations correlation equations have been derived for the local compaction rate, and the following influential parameters identified:

- the surface temperature of the deposit/substrate (as a scale for the thermal state of the surface),
- the impinging momentum of the droplets, and
- the droplet solidification ratio (or liquid contents).

Buchholz (2002) found that the compaction rate is not very sensitive to variations in the impact angle of the particles.

The main behaviour during compaction is the impact of the droplet and the occurrence of drop deformation and fragmentation. During impact of fluid droplets normal to solid/liquid surfaces (without phase change or solidification), three possible regimes (see principle sketch in Figure 6.2) may be identified (see, for example, Armster *et al.* (2002), Rein (1993)):

(1) partial or complete rebounding or repelling of the droplet,
(2) splashing of the droplet,
(3) partial fragmentation of the impinging droplet, or the liquid impact surface, with secondary droplet formation.

Simulation of drop-impact processes increases our understanding of drop-impact physics and enables the derivation of impact models. The principal numerical methods used to investigate drop impingement process are based on the early work of Harlow and Shannon (1967). The numerical analyses increase in complexity from (1) to (3) in the above list.

As a first approach, the dependence of spreading behaviour and splashing limit on the main influencing parameters needs to be analysed. In this context, the numerical approach discussed in Section 4.1.2, for analysis of the behaviour and movement of free liquid surfaces,

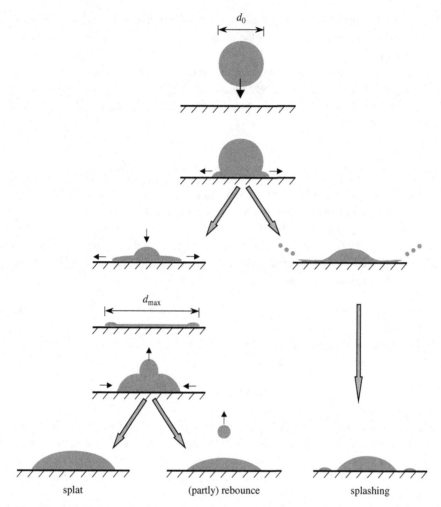

Fig. 6.2 Liquid droplet impact on a solid surface: different results

may be used, taking into account the surface tension algorithm of Brackbill *et al.* (1992). In addition, the contact angle or three-phase angle at the liquid (*l*)/solid (*s*)/gas (*g*) contact point or line is incorporated. The static contact angle, defined in Figure 6.3, demonstrates the case of equilibrium between phases, and from Young's formula is:

$$\cos(\alpha) = \cos(180° - \theta) = \frac{\sigma_{s,g} - \sigma_{s,l}}{\sigma_{l,g}}. \tag{6.2}$$

For water-based or organic solutions, which have been much studied in the literature, during impact, the contact angle $0° < \theta < 90°$ determines the hydrophobic behaviour of the fluid at the surface; while for $90° < \theta < 180°$, the behaviour is hydrophilic. For complete wetting of the solid surface by the liquid, the static contact angle is $180°$.

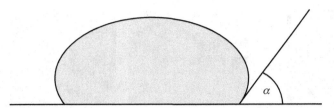

Fig. 6.3 Definition of static contact angle

6.1.1 Drop deformation

Figure 6.4 shows a simulated low-speed drop impact sequence and the deformation process for a liquid (non-solidifying) droplet on a solid surface, based on the boundary conditions, listed in Table 6.1. Two different droplet fluids are studied: water and tin melt (where at first no heat transfer and no solidification of the droplet material is assumed, the impact process is isotherm).

For the relevant characteristic numbers of the droplet impact process, from the liquid properties and the impact velocities, the values of Reynolds number Re $=$ 5000 and Weber number We $=$ 137 for the water droplet, and Re $=$ 13 318 and We $=$ 72 for the tin drop, may be calculated.

From the calculation of the impact process in Figure 6.4 four distinct temporal stages of drop deformation can be observed:

(1) $t < 3$ ms: the droplet splashes and flattens on the solid surface. In this stage, mainly the kinetic energy of the impacting droplet is transformed into surface energy.
(2) $3 < t < 5$ ms: a ring-shaped rim at the circumference of the droplet is observed. In this stage, the rate of growth of the splat diameter decreases. The inner area of the droplet is drawn out to a thin film.
(3) $5 < t < 25$ ms: recoiling of the fluid due to the action of surface tension is seen. This behaviour may, in extreme cases, lead to complete detachment of the droplet (rebound) from the surface (see, for example, Ford and Furmidge (1967)).
(4) $t > 25$ ms: the deformed droplet may oscillate around its steady state until the final shape of the stationary droplet on the surface is achieved (not shown in the present sequence).

Figure 6.5 compares the results from the above numerical simulation with the experimental results of Berg and Ulrich (1997) for an unsolidified tin droplet impacting at 90° to the solid surface. These experimental results have been achieved for identical process conditions and melt properties as in the simulation. The figure shows the time-dependent behaviour of the maximum splat diameter of the droplet during impact. The measured results indicate a more rapid flattening of the droplet in the first impingement phase. The maximum achieved droplet diameter in the steady state is of comparable order. In the experiment, the tin droplet solidifies after splashing. Therefore, the splashing kinetics is far faster than the solidification kinetics in this case.

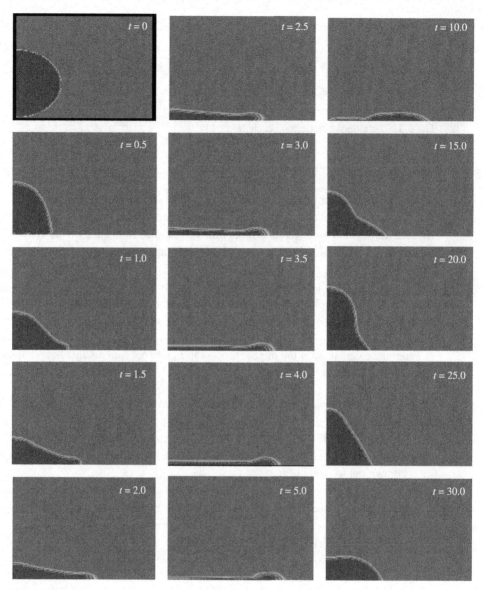

Fig. 6.4 Impact of a water drop: $d = 2.5$ mm, $u = 2$ m/s, time scale in ms

A widely used model for estimation of the maximum splat diameter of impinging metal droplets has been derived by Madejski (1976). Accounting for viscosity and surface tension effects, as well as for the solidification behaviour of droplets, theoretical derivation of the model is based on two-dimensional radial flow. Piecewise fitting of the result obtains, for the general case (without solidification), in the range of relevant characteristic numbers

Table 6.1 *Data for droplet impact calculations*

	density ρ [kg/m³]	dynamic viscosity μ [kg/m s]	surface tension σ [N/m]	contact angle [deg.]	velocity u [m/s]	particle diameter d [μm]
H_2O	1000	0.001	0.0725	115	2.0	2500
Sn	7000	0.001 85	0.544	125	1.6	2200

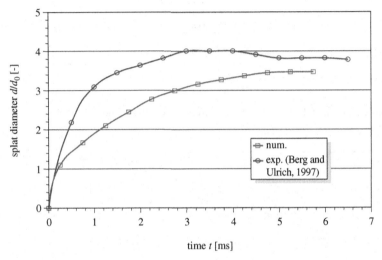

Fig. 6.5 Impact of a tin drop: splat diameter, $d = 2.2$ mm, $u = 1.6$ m/s

$Re > 100$ and $We > 100$:

$$\frac{3d_{max}/d_0}{We} + \frac{1}{Re}\left(\frac{d_{max}/d_0}{1.2941}\right)^5 = 1. \tag{6.3}$$

From this equation, the case of the splashing tin droplet in Figure 6.3 has been recalculated. The result indicates a value of the maximum splat diameter of $d_{max}/d_0 = 4.8$, a somewhat higher value than found in either the experiment or the simulation.

   In another empirical model, based on the work of Scheller and Bousfield (1995), the maximum splat diameter of impinging droplets for different liquids has been correlated in terms of the Reynolds number and the Ohnesorge number as:

$$\frac{d_{max}}{d_0} = 0.61(Re^2\, Oh)^{0.166}. \tag{6.4}$$

Here the main parameters determining the drop impact are the fluid properties of the imping-ing droplet, the drop size and the drop-impact velocity. Evaluating this correlation together with the measurement results of Berg and Ulrich (1997) for the impact and deformation of tin, water and glycerol droplets, plus the simulation results for the drop impact (numerical),

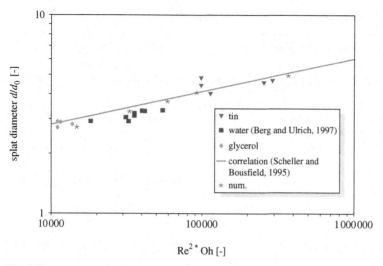

Fig. 6.6 Maximum splat diameter (correlation from Scheller and Bousfield, 1995)

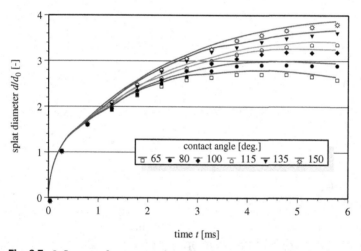

Fig. 6.7 Influence of contact angle on droplet spreading during impact

Figure 6.6 indicates the relevance of Eq. (6.4) for model derivations in agreement with experimental and simulation results.

The influence of static contact angle on the droplet splashing process has been examined and is illustrated in Figure 6.7 for normal (i.e. 90°) drop impact and the behaviour of the time-dependent drop splash diameter. By increasing the static contact angle, the droplet splashing behaviour is mainly affected in the second stage of the splashing process. Here the maximum drop splash diameter increases with increasing values of the contact angle. The same behaviour has been identified by Scheller and Bousfield (1995), but at a lower relevance level, who found that the maximum splash diameter changed by as much as 10% when the contact angle was in the range of 85 to 145°. Scheller and Bousfield also found

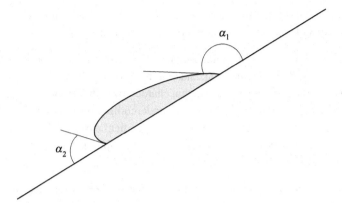

Fig. 6.8 Hysteresis of the contact angle

that for higher Reynolds numbers, the surface roughness of the solid is of minor importance to the maximum splat diameter.

The role of contact angle in the droplet-impact process still needs further evaluation. Instead of taking just the static contact angle, for rapid wetting problems the more relevant dynamic contact angle needs to be incorporated (when data are available, especially for metal melts). Also hysteresis of the contact angle during droplet spreading and recoiling (see Figure 6.8 for an example of a drop on an inclined surface) is to be used. Once again, it is the lag of physical data and properties for metal melts in the high-temperature range that has prevented the inclusion of these important effects into numerical models up to now.

At higher impact velocity, the droplet may disintegrate during impact and satellite droplets may form. The boundary condition between complete droplet deposition and fragmentation in this splashing regime has been derived by Walzel (1980) and has been tested for molten metal droplet impact by Berg and Ulrich (1997): see below.

Direct numerical calculations of this splashing behaviour have been performed, for example, at the University of Stuttgart. Results and animation of the simulated drop impact and splashing process on rigid surfaces and liquid films may be found at: www.uni-stuttgart.de/UNIuser/itlr/gallery.html (see Useful web pages, p. 269). Here simulated animations of binary drop collision processes are also to be found. Direct numerical simulations require high computer power due to the fine grids used in transient three-dimensional calculation (see below).

6.1.2   Droplet solidification during impact

Numerical investigations of metal drop-impact processes, together with a simultaneous calculation of phase change behaviour and the solidification process, have been performed by Delplanque *et al.* (1996), Fukai *et al.* (1998), Liu *et al.* (1993, 1994a), Trapaga *et al.* (1992) and Waldvogel and Poulikakos (1997).

Starting from methods discussed above for evaluation of the fluid dynamics of drops and phase boundaries in these contributions, a phase change model has been derived and implemented into numerical simulation codes. Initially, one-dimensional solidification modelling has been investigated. The models are based on two basic assumptions:

(1) It is assumed that the moving solidification front in the material strictly separates regions of fully solidified and fully liquid material, and that the distribution of this solidification front is described by that particular isotherm that characterizes the temperature of solidification (no solidification temperature range, no undercooling prior to solidification).
(2) The main direction of movement of the heat flux is perpendicular to the substrate surface, from the droplet towards the substrate. This allows the assumption of a one-dimensional heat transport mechanism.

Based on these assumptions, analysis of the phase change process is reduced to a one-dimensional problem of a moving solidification front. The solidification front may only move perpendicular to the substrate and only away from it. The thickness of the already solidified layer s in the droplet is described by one-dimensional solution of the Stefan solidification problem (Hill, 1987; Madejski, 1976; San Marchi et al., 1993):

$$s = 2\lambda_e \sqrt{a_s(t - t_0)}. \tag{6.5}$$

In this equation, t_0 is the initial time of origin of the solidification process. The solidification coefficient λ_e is derived from a heat balance at the solidification front as:

$$\lambda_e = \frac{1}{\sqrt{\pi}} \left[\frac{T_s^*}{\mathrm{erf}(\lambda_e) \exp\left(\lambda_e^2\right)} - \frac{T_l^* \sqrt{a_l/a_s}}{\mathrm{erf}(\lambda_e \sqrt{a_s/a_l}) \exp\left(\lambda_e^2 a_s/a_l\right)} \right]. \tag{6.6}$$

Here the dimensionless temperatures (Stefan numbers) of the liquid material (index s – solid) and the liquid material (index l – liquid) are:

$$T_s^* = \frac{\lambda_s(T_m - T_s)}{a_s \rho_s \Delta h_{sl}}, \quad T_l^* = \frac{\lambda_l(T_l - T_m)}{a_l \rho_l \Delta h_{sl}}. \tag{6.7}$$

Within the numerical realization of this model in a simulation code, the already solidified layers within the droplet are rejected from the solution area where the fluid movement is calculated. These cells are assumed to be immobile (i.e. wall cells).

Based on this, Delplanque et al. (1996) and Liu et al. (1994a) developed a model for the formation of micropores within thermal spray or spray forming processes. As an example, the normal (i.e. 90°) drop-impact process of a 30 μm tungsten droplet at an impact velocity of 400 m/s on a rigid surface has been calculated. These impact conditions are related to typical thermal spray process properties. The initial temperature of the droplet is 100 K above the melting point of 3650 K, the substrate has an initial temperature of 1500 K. The simulated drop deformation process during impact and the establishment of the solidification area in the droplet are illustrated in Figure 6.9.

At the start of the deformation process, i.e. at the initial point of impact between the drop and the surface, a thin solid layer is formed. Due to the high radial velocities of the liquid layer just above the solidified area, shortly after contact ($t = 0.045$ μs), a radial liquid jet is formed at the outer edge of the tip which overshoots the underlying solid region. Due to the circumferential symmetry assumed in the calculation, this jet has the geometry of a radially

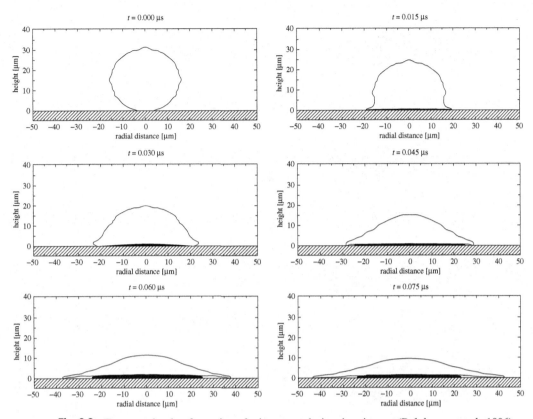

Fig. 6.9 Tungsten droplet: formation of micropores during drop impact (Delplanque *et al.*, 1996)

spreading liquid sheet. This jet detaches from the surface and, due to inertia and gravity, reattaches to the substrate surface some distance from the detachment point of the droplet. A pore remains below the liquid bridge formed by this jet. If the bridging layer solidifies before the pore is filled, a pore is formed in the bottom surface of the spread droplet. If the pore is filled by some liquid from above, a crater at the top surface of the droplet may be formed and remains after solidification. This crater may act as an initial pore source if a subsequent droplet during impact rapidly flows across the crater in a radial direction. The one-dimensional character of the solidification front model limits the usefulness of such microprocesses to the impact of solidifying melt drops. Multidimensional phase change models therefore need to be developed, as has been done, for example, by Delplanque *et al.* (1996). This model has also been derived and applied for molten metal drop impact processes.

6.1.3   Secondary atomization during impact

Not only during impact on liquid pools or films, but also at specific regions during impact on solid surfaces, a fluid lamella is formed at the rim of the droplet which may be unstable

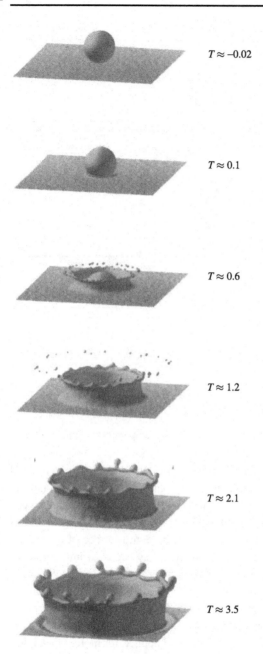

$T \approx -0.02$

$T \approx 0.1$

$T \approx 0.6$

$T \approx 1.2$

$T \approx 2.1$

$T \approx 3.5$

Fig. 6.10 Three-dimensional simulation of droplet impact onto a shallow liquid layer: We $= 250$, Re $\sim 10\,000$, $h/d = 0.12$, $T = tv_p/d_p$ (Frohn and Roth, 2000)

and therefore may disintegrate due to capillary effects into smaller (secondary) droplets. Because of the three-dimensional character of this process and the ratio between the main drop and tiny satellite drops (where very fine grids are needed in the whole solution domain), full simulations of the secondary spray formation mechanism have not been frequently published. The normal drop impingement process on a rigid wall with secondary drop formation has been investigated by Rieber and Frohn (1998). The calculation is based on a VOF method for free surface analysis (see Section 4.1.2). A grid system consisting of 7.1×10^6 cells has been used in this computation. To achieve physically realistic results, the liquid lamella formed at the rim of the spreading droplet must be artificially excited in the simulation. Without excitation the lamella disintegrates too, but due to numerical instabilities caused by too coarse a grid resolution and not as a result of physical instability processes.

As an example of the simulation of a three-dimensional droplet splashing into a shallow liquid pool, involving secondary fragmentation, Figure 6.10 illustrates the time sequence for disintegration of the splashing lamella after impact: We = 250, Re $\sim$ 10 000 and the dimensionless thickness of the liquid pool $h/d = 0.12$. A 320^3 grid at 2.553 integration time steps has been used for the simulation. Only one-quarter of the visualized splash was really simulated in the numerical calculations. The drop impact results in splashing, in agreement with experimental observations. The formation of a characteristic crown with finger-ejecting small droplets is to be seen.

Experimental and numerical investigations of drop impact and splashing phenomena on solid (dry or wetted) walls have also been investigated by Mundo (1996). Dependent on the drop impact Reynolds number and the Ohnesorge number of the liquid, based on single-drop impingement experiments, the splashing limit has been identified. In Mundo (1996), the splashing limit has been given by means of a characteristic number $K = \text{Oh Re}^{1.25}$. In addition, the droplet-size distribution in the secondary spray from droplet splashing has been investigated by means of a PDA device. From this analysis, a particle impact model is derived that can be included in spray impact simulation codes based on a Eulerian/Lagrangian approach.

An extension of the VOF approach for analysis of the liquid/gas interface has been introduced by Bussmann et al. (1999). The model allows the simulation of inherently three-dimensional structures during droplet impact, e.g. for inclined impact, impact onto structured substrates (rough surface) and splashing effects. Figure 6.11 shows the three-dimensional impact of a molten metal (tin) droplet onto a solid surface (Bussmann et al., 2000). The numerical results are in good agreement with the experimental in terms of spreading and the splashing region. The number of satellite droplets ejecting from the rim of the droplet is captured.

The impact and solidification of tin droplets on a flat steel plate was studied by Pasandideh-Fard et al. (1998) using experimental and numerical simulations. In the experiments, tin droplets (2.1 mm in diameter) were formed and dropped onto a stainless steel surface whose temperature was varied from 25 to 240 °C. The impact process of droplets was photographed, and evolution of droplet spread diameter and liquid–solid contact angle were measured

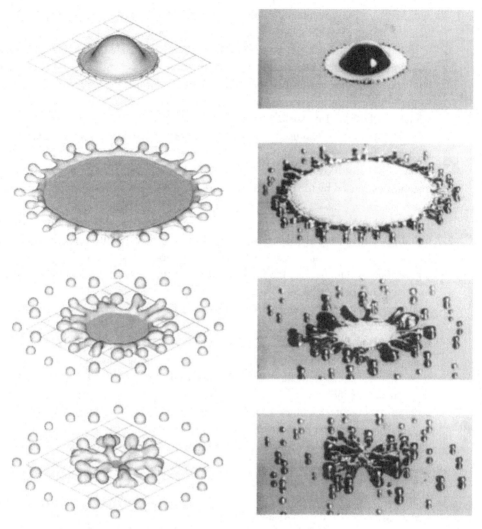

Fig. 6.11 Three-dimensional impact of a tin droplet onto a solid surface: comparison between simulation (left) and experiment (right) (Bussmann *et al.*, 2000)

from photographs. The measured values of the liquid–solid contact angle were used as a boundary condition for the numerical VOF model. An example of the simulation results is shown in Figure 6.12 for the impact of a tin drop onto a substrate initially at 150 °C. The drop shape, and temperature evolution in the drop and the underlying substrate are to be seen. The heat transfer coefficient at the droplet–substrate interface was estimated by matching numerical predictions of the variation in substrate temperature with measurements. Comparison of computer-generated images of impacting droplets with photographs showed that the numerical model correctly predicted droplet shapes during impact for simultaneous spreading and solidification. From these results an empirical correlation with the maximum

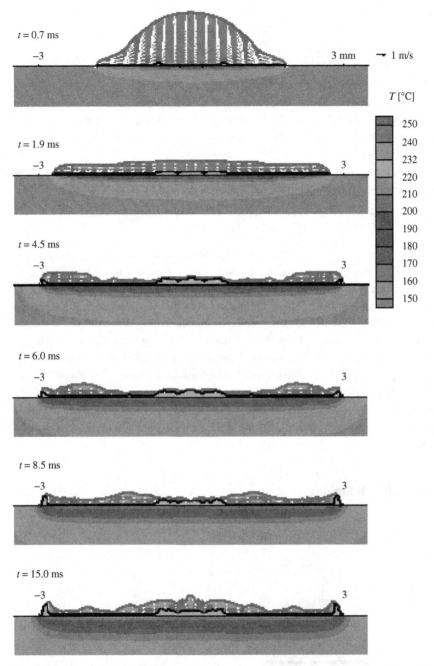

Fig. 6.12 Drop impact simulation of tin ($d_p = 2.1$ mm) onto a stainless steel substrate (Pasandideh-Fard *et al.*, 1998)

spread diameter of a freezing droplet has been derived:

$$\frac{d_{max}}{d_0} = \sqrt{\frac{We + 12}{3(1 - \cos \Theta) + 4(We/\sqrt{Re}) + We\sqrt{(3Ste)/(4Pe)}}}. \tag{6.8}$$

Here the relevant kinetic and thermal numbers for droplet impact are introduced, as the Weber number We, the Reynolds number Re, the Stefan number Ste and the Peclet number Pe, as well as the liquid–solid contact angle Θ. The magnitude of the term $\sqrt{(3Ste)/(4Pe)}$ determines whether solidification influences the droplet spreading process or just the kinetic impact energy.

From least square analysis of the observed outcome of molten metal drop impingement experiments, a splashing number Z^* has been derived by Berg (1999) as:

$$Z^* = 9.48 \frac{Re^{0.177} \, We^{0.224} \, Pe_T^{0.115} \, Pe_S^{0.021} \, Ec^{0.108} \, \beta^{0.335}}{Ste^{0.116} \, Nu^{0.103} \, (1 - \cos \Theta)^{0.08}} = 825. \tag{6.9}$$

Droplets impacting at $Z^* < 825$ will only spread and completely deposit, droplets at $Z^* > 835$ will partly fragment during impact. The correlation has been derived from drop experiments with tin, lead, copper, aluminium and steel, impinging perpendicularly to a stainless steel substrate. The correlation is valid in the investigated range of Reynolds numbers $7500 < Re < 135\,000$ and Weber numbers $50 < We < 3700$.

In summary, adequate modelling of particle and drop impact processes is essential for numerical description of impact-orientated spray processes within coating or spray forming applications. In the latter, the ratio of back splashing to compacting droplet mass is also of special interest, as it determines the compaction efficiency of the spray forming process. Also the compaction rate or efficiency is an important input parameter for modelling the geometry of growing preforms during spray forming, which will be introduced in the next section.

6.2   Geometric modelling

The possibility of producing near-net shaped preforms is one of the most important advantages of the spray forming process. Flat products (sheets), tubular products and rings, and massive volumetric products such as billets, are the most often produced geometries for industrial application. The spray forming of more complicated preforms and geometries is the subject of much research and development or testing. A necessary condition for the adaptation of spray forming process to new geometries is the ability to predict resulting preform geometries from process conditions and material parameters. Also the planning of operational parameters in advance and process control during operation of a spray forming facility require information on the transient shape of the ideal spray formed preform in order to control the process properly. The geometry of spray formed products depends on the mass flux distribution in the spray, the substrate geometry and movement, and the compaction rate or efficiency.

Continuous shape modelling of preforms during spray forming, which defines the spray as a continuous mass flux released from a source, has been developed by Frigaard (1994, 1997, 2000). Formal description of the spray is given in terms of its resulting shape during application onto a surface with the source (spray) axis pointing onto the substrate. This function is of Gaussian type; the corresponding mean and standard deviation need to be obtained from experiments. The advantage of this model is its great flexibility in dealing with more complicated geometric arrangements of substrate and spray. However, two important problems must be solved. First, in every time step, knowledge of the visibility of any points on the surface, as seen from the spray origin, is required. All points within the shadow regions will not advance in spray during such a time step until they become visible. The algorithm calculating the visibility function is required for every time step and has a strong impact on the overall computational time. Second, the position of the surface after each discrete time step must be tracked, followed by reconnection of the surface segments to define the new surface position. This procedure is controlled by a so-called small-distance parameter and may lead to merging of the closest surface parts, resulting in the formation of macropores (Djuric $et\ al.$, 1999).

The geometric correlation for estimating the local growth rate g (increase in height per time unit) is:

$$\dot{g} = \frac{\dot{m}_{spray}}{\rho_p} \frac{f_{sh} \cos \beta}{E_{por}} k_p, \tag{6.10}$$

and is related to growth perpendicular to the surface. The angle β is that between the surface normal on the deposit contour and the particle impingement vector. If the growth is to be determined in the direction of the spray centre-line, then $\cos \beta = 1$. The mass flux distribution in the spray is related to the theoretical material density and the local compaction rate (or sticking efficiency) k_p. The shadow factor f_{sh} is a visibility factor which may have values between $f_{sh} = 1$ (visible area on the deposit from the viewpoint of the spray impact normal, related to the particle trajectory vector before impingement) and $f_{sh} = 0$ (a point within the shadow area of another deposit area, related to the particle trajectory vector before impingement). A porosity factor E_{por} is introduced which describes the difference between the local theoretical material density (density of droplets ρ_p) and the actual resulting material density in the deposit including pore formation (where the definition of the local porosity Ψ is $E_{por} = 1 - \Psi$). The compaction rate is either assumed to be constant across the whole compaction area or is a function of the local position on the deposit surface, or

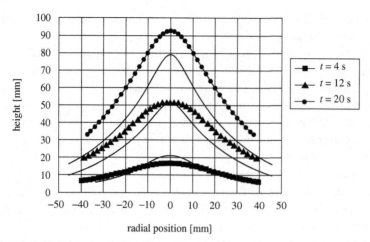

Fig. 6.13 Calculated deposit contour (Kramer, 1997), comparison between experimental values (line): constant compaction rate $k = 0.8$

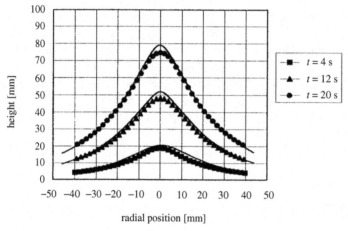

Fig. 6.14 Calculated deposit contour (Kramer, 1997), comparison with experimental values (line): variable compaction rate

is based on other models dependent on position and local properties (such as the surface temperature, see Section 6.1 (Buchholz, 2002; Kramer *et al.*, 1997).

In Figures 6.13 and 6.14, comparison between measured transient deposit contours and model-based calculated deposit contours is illustrated. In the first figure it is assumed that the compaction rate is constant over the entire surface of the deposit ($k_p = 0.8$). In the second figure, the compaction rate has been taken from an empirical function based on measurement results. In both figures, spray forming by means of a stationary atomizer spraying perpendicular on a cylindrical substrate is assumed. For such operational parameters, a Gaussian-shaped deposit will result, as can be seen in the figures. The results have been taken from an investigation of Kramer (1997). The calculation with a temporal and

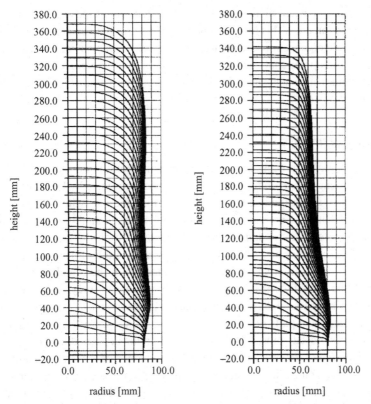

Fig. 6.15 Calculated geometry of a growing billet in the spray phase (Kienzler and Schröder, 1997)

spatial constant compaction rate yields, after a certain time, relevant deviations from the real geometry of the measured deposit. This deficit may be related to the fact that the compaction rate in the initial phase of the spray forming process for Gaussian-shaped geometries depends strongly on time and location (radius). By using a quite high compaction rate of $k_p = 0.8$ (80% compacted material, which is quite high for this geometry but may be higher for other geometries and materials) the geometry is reflected correctly at the initial stage of the process; but at later stages, when the compaction rate is decreased, growth of the deposit is overpredicted. By using a variable compaction rate, as in Figure 6.14, the calculated growth and deposit geometry are in quite good agreement with experimental data.

The growth of a cylindrical billet during spray forming is illustrated in Figure 6.15 for a typical spraying time of 350 s. This type of billet is usually produced by sidewise spraying onto a rotating cylindrical substrate that is withdrawn downward to maintain a constant distance between atomizer and impingement area (to keep the spray parameters constant during impact). The figure shows calculated contour results with and without the use of a simple compaction rate model. The withdrawal speed of the substrate in this case was 1 mm/s. The growth behaviour has been calculated and two different growth phases have been obtained. After the initial phase, where the billet contour changes steadily (billet growth rate), in the later phase, after a spraying time of $t > 100$ s, growth is stationary;

the billet grows at a constant diameter and obtains a constant shape and contour at its tip within the compaction area. This result has been achieved for both calculations, though the calculation from the compaction rate model yields a pure convex outer surface of the billet (as intended). If the operational conditions are kept constant, the billet will grow straight upwards afterwards with a constant contour at arbitrary times (limited just by the technical restrictions of the process).

6.2.1   Form-filling spray process

Modelling of the spraying process of a liquid metal, which solidifies on an arbitrary surface, and of the subsequent growth of the deposit has also been presented by Djuric *et al.* (1999). The continuous two-dimensional model developed defines the source involved in the spraying process; the kinetics of points on an arbitrary substrate, including their visibility and the spray's sticking efficiency; and redefines a new surface on the completion of every iteration step. The surface is described by a multivalued function of its spatial coordinates (which does not cross itself at any point) at every moment of the calculation. The source is a function of the mass flux of the material arriving from the spray, and its distribution and movement in space is taken from corresponding experiments. A merging procedure is developed to reconnect parts of a curve, which become too close during spraying. This procedure produces closed voids, called macropores, in the deposited material, defined as a continuous function of the spray parameters. The influence of the spraying angle, as the most important spray parameter, on shape evolution during filling of a (two-dimensional) notch with slightly sloped walls is discussed. A porosity distribution function is calculated for every set of input parameters, and its relationship to shape evolution emphasized. The computational procedure consists of four steps:

- description of the curve (surface) at any moment,
- calculation of the visibility function and other geometric parameters,
- curve evolution because of the impacting mass, and
- definition of a new curve at the end of every time step.

Some global results have been achieved from these calculations, such as:

- the formation and distribution of macropores caused by the merging of closest curve segments, and
- the microporosity distribution in the deposited material.

An empirical local microporosity function p has been modelled as a function of the impact angle Θ_{col}, defined as the angle between the radius vector of any point on the curve and the tangent vector at the same point. This function has been taken from the experimental results of Smith *et al.* (1994) in the form of a best-fit polynomial approximation:

$$p(\Theta_{col}) = a_0 + a_1\Theta_{col} + a_2\Theta_{col}^2 + a_3\Theta_{col}^3 + a_4\Theta_{col}^4, \tag{6.11}$$

where the coefficients are $a_0 = 0.622$, $a_1 = -1.749$, $a_2 = 2.093$, $a_3 = -1.115$ and $a_4 = 0.222$.

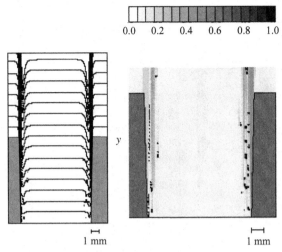

Fig. 6.16 Simulated notch filling geometry for $\alpha_{\text{spray}} = 0°$ and $k_p = 1$. Left: shape evolution. Right: microporosity function (Djuric *et al.*, 1999)

The model has been applied to the two-dimensional notch filling case using different spray angles α_{spray} in the range 0° to 40°, where the source (spray) is moved along a horizontal line at a certain distance from the top of the initial shape and with a prescribed velocity. The mass flux distribution of the source has been taken from a standard experiment (footprint experiment). For the sticking efficiency k_p, either 100% (complete compaction) or an empirical correlation, also approximating the experiments in Smith *et al.* (1994) as a function of the impact angle, have been used:

$$k_p(\Theta_{\text{col}}) = b_0 + b_1 \Theta_{\text{col}} + b_2 \Theta_{\text{col}}^2, \tag{6.12}$$

where the coefficients are $b_0 = 0.339$, $b_1 = 0.875$, $b_2 = -0.292$.

The simulation results for two different spray inclination angles are illustrated in Figure 6.16 for $\alpha_{\text{spray}} = 0°$ and $k_p = 1$, and in Figure 6.17 for $\alpha_{\text{spray}} = 10°$ and $k_p = 1$ (Djuric *et al.*, 1999). On the left, the spray formed shape evolution while filling the notch in the spray phase is seen; while on the right, the resulting microporosity and macropore formation at the end of the spray time is seen. For the various spray angles discussed, different macroporosity (bridging) and microporosity distributions are calculated. Some final spray formed profiles have been compared to experimental profiles, which show good agreement. From these calculations, the process parameters influencing spray forming may be derived.

In a further development of this model, Djuric and Grant (2001) included submodels for splashing and redeposition of droplets during impact onto the deposit (still in two dimensions) for spray forming of a notch geometry. By including or excluding droplet-splashing effects, it has been possible to analyse the importance of droplet redeposition both in simulations and experiments. In the case of splashing, the different microsource functions assumed resulted in slightly different, but geometrically similar, preform shapes. Modelling of these functions was based on: continuous splashing/reemission processes, a simple point

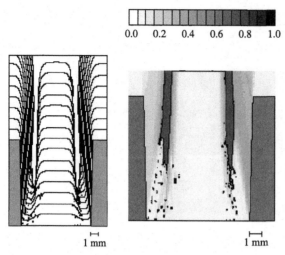

Fig. 6.17 Simulated notch filling geometry for $\alpha_{\text{spray}} = 10°$ and $k_p = 1$. Left: shape evolution. Right: microporosity function (Djuric *et al.*, 1999)

source, qualitative analysis of splashing events observed during experimentation by high-speed imaging, a total mass preservation constraint, and some theoretical and experimental results of single-droplet impact onto hard surfaces. From the results of this model, it was possible to predict the impact regions where direct spraying dominated deposit growth and where subsequent redeposition has a major influence on deposit growth and the final shape. In order to address the remaining deficiencies in the model, resulting from comparison to experimental values, three-dimensional effects need to be taken into account.

6.2.2   Computer-hardware-tailored geometric simulation

A general geometric model describing the evolution of the deposit shape, including the effect of multiple atomizers, as well as controlled atomizer scanning within a gear-driven device, has been developed by Markus and Fritsching (2003).

In a spray forming application with two independent atomizers for melt disintegration, one or both nozzles may scan periodically around a fixed axis (see Figure 6.18). The eccentricity e is the distance between the central point of the substrate and the spray cone axis at its zero position. The substrate rotates and is translated away from the atomizer in order to maintain a constant distance from the surface to the atomizer that compensates for growth of the deposit. The deposit is described in Cartesian coordinates. A structured grid describes the deposit surface initially containing $n \times m$ nodes (each node located at the centre of a cell). While the deposit is growing, the position of the nodes will move. In the case of small cells, a certain number may be combined to form a single new cell. In the case of coarse cells, these may be subdivided to form new, smaller cells. During spray deposition, the cells on the surface may deform and the grid may be properly smoothed in such a way that the nodes are equally distributed on the deposit surface.

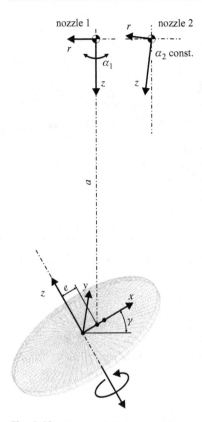

Fig. 6.18 Sketch of the geometric arrangement of nozzles and substrate for billet production by multiple atomization nozzles

Figure 6.19 illustrates some resulting surface meshes for different spray formed deposits. The geometries of a general, noncircular billet (*a*), a ring (*b*) and tube (*c*) are illustrated, respectively. Where periodic agreement between substrate rotation and the atomizer nozzle scanning frequency occurs, an irregular shape may be produced as in Figure 6.19(*a*).

The spray is described within a cylindrical coordinate system with its origin at the nozzle. In order to simulate growth of the deposit, the local distribution of the melt mass flux in the spray $\dot{m}_A$ needs to be prescribed, and can be derived from the experimental correlation in Uhlenwinkel (1992). For a certain material and a constant melt and gas mass flow rate, the radial distribution of the mass flux density at a certain position z can be described as a Gaussian-shaped function:

$$\dot{m}_A(r, z) = \dot{m}_{A,\max} e^{\ln(0.5)(\frac{r}{r_{0.5}})^{k_1}}, \tag{6.13}$$

with the maximum mass flux $\dot{m}_{A,\max}$ at the centre-line of the spray at $r = 0$. The skewness of the function is described by an empirical constant k_1, which was found from experiment to be $k_1 = 1.4$. At the half-width radius $r_{0.5}$, the local mass flux density is decreased to half

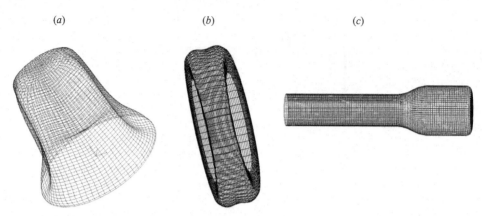

Fig. 6.19 Examples of surface grids for: (a) billet, (b) ring and (c) tube geometry simulation

the maximum mass flux value $\dot{m}_{A,\max}$ at the centre-line of the spray. The following equation shows the correlation between $\dot{m}_{A,\max}$ and $r_{0.5}$:

$$r_{0.5} = \sqrt{\frac{k_2 \dot{M}_{\text{melt}}}{\dot{m}_{A,\max}}}.$$ (6.14)

The width of the spray cone is assumed to increase with increasing distance from the nozzle in a non-linear way:

$$r_{0.5} = r_{0.5,\text{ref}} \left(\frac{z}{z_{\text{ref}}}\right)^{\frac{k_{\text{ref}}}{2}},$$ (6.15)

where the parameter k_{ref} for a reference distance z_{ref}, 500 mm from the nozzle, is about 1.5 (Buchholz, 2002). A constant sticking efficiency of the spray particles on the surface is assumed.

The visibility of surface points is calculated using a computer graphics algorithm. The problem of checking whether an object is visible or not, is a common task within computer graphics. Two methods may be used to check the visibility:

(1) back face culling, and
(2) the Z-buffer algorithm.

A simple and fast way of checking visibility is to use back face culling. It just checks if the normal onto the cell surface can be seen in the direction of the viewpoint or not. It can be calculated by simple vector algebra:

$$\mathbf{n} \times \mathbf{v} \quad \frac{\geq 0 \; : \; \text{invisible}}{< 0 \; : \; \text{visible}},$$ (6.16)

with $\mathbf{n}$ the direction normal to the vector at the cell surface and $\mathbf{v}$ the direction towards the viewpoint. For completely convex bodies, this method results in correct visibility functions, but for complex shapes some problems occur. Another way to check the visibility of surface

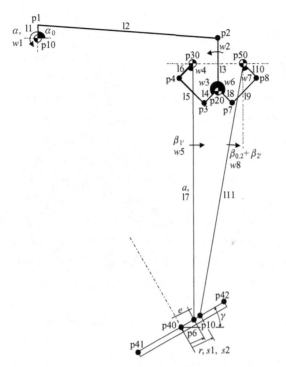

Fig. 6.20   Illustration of gear mechanism for control of two atomizer nozzles

points completely using computer graphics is to employ the Z-buffer algorithm. This algorithm typically forms part of the computer graphics hardware and is therefore very fast. All objects will be flushed to the graphics card memory, where each cell is assigned, for example, to a unique colour. By reading the colours directly from the graphics card memory, the visibility of each cell can be checked in a quick and simple way. The necessary graphics programming is simplified by the availability of general graphics software packages such as OpenGL.

The gear mechanism illustrated in Figure 6.20, which allows for the general scanning movement of the atomizers, consists of a number of simple devices. Kinematic analysis of the modular gear system is carried out independently.

Figure 6.21 compares the geometry of a simulated billet with that produced during spraying using conventional process parameters. In this experiment a relatively small melt orifice has been tested, with a diameter of just 1.6 mm, spraying at a small distance between atomizer and substrate. In this case, the spray parameters for the resulting very low melt flow rate have been extrapolated from the mass flux distribution function of Uhlenwinkel (1992). The resulting shape of the simulation agrees well with experiment.

A billet formed from the simultaneous spraying of two atomizers is shown in Figure 6.22. It has a weight of 112 kg, a height of 335 mm and a diameter of 249 mm. The simulation gives 335×269 mm for the dimensions of the billet.

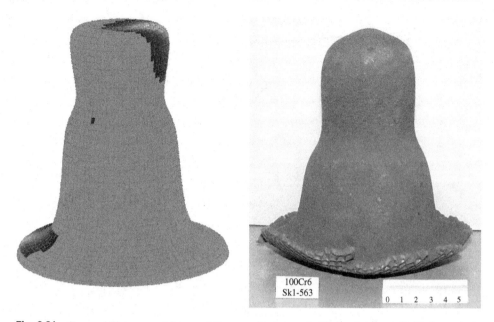

Fig. 6.21 Copper billet sprayed from a 1.6 mm melt stream. Left: simulated result. Right: experimental result

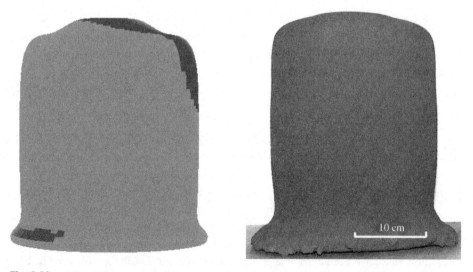

Fig. 6.22 CuSn$_6$ billet sprayed with two atomizers

6.3   Billet cooling

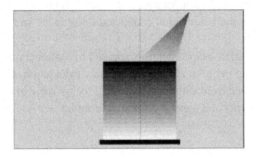

The cooling conditions experienced by the metal during spray forming significantly determine the resulting properties of the material. As spray formed metals are widely assumed
to be superior compared to conventionally manufactured materials, analysis of the cooling
conditions is a key process for understanding and optimization of spray forming processes.
In the following section, thermal analysis and simulation of the solidification and cooling
behaviour of billets during two stages of spray forming, namely the growth process (during
the spray phase) and the subsequent cooling process, will be discussed. Here, modelling of
the behaviour of the deposit, together with its underlying substrate, and including all necessary submodels and boundary conditions, is introduced. Using a numerical algorithm, the
process parameters, which are influenced by variable boundary conditions during the spray
run, are identified and the thermal state of the deposit, in terms of the local temperature and
solidification state, is derived.

6.3.1   Method

For analysis of the thermal conditions in the deposit and substrate, the heat conduction
(Fourier) equation including a source term that accounts for latent heat release, in differential
form, is discretized by means of finite differences in the solution domain. Fitting a grid
structure to the actual contour of the deposit during coordinate transformation simplifies
the mathematical description, as well as numerical handling of the boundary conditions at
curves. Here the grid lines and discretization points are directly aligned with the contour
(boundary fitted coordinates, BFC), thereby increasing the accuracy of the calculation.

6.3.2   Assumptions

A whole series of assumptions and simplifying boundary conditions need to be made for
initial calculation of the thermal state of a Gaussian preform:

- Material properties such as density ρ, heat capacity c_p and heat conductivity λ are independent of temperature.

- The state of the impinging spray does not change with time nor with the height of the deposit during the growth phase.
- To model the phase change behaviour, heterogeneous nucleation is assumed. This describes the solidification process while the temperature remains constant until the local latent heat amount is released from the volume and conducted towards the surrounding finite volume elements.
- The first calculation discusses Gaussian-shaped deposits sprayed by stationary atomization perpendicular to a non-moving cylindrical substrate. In this calculation, the mass flux distribution in the spray is derived from Uhlenwinkel (1992) for spraying of different metals. This can be expressed in terms of a general empirical formula:

$$\frac{\dot{m}}{\dot{m}_c} = e^{-k_1(\frac{r}{r_{0.5}})^{k_2}} \tag{6.17}$$

- where $\dot{m}_c$ is the maximum value of the mass flux distribution at the centre-line of the spray and $r_{0.5}$ is the half-width of the mass flux distribution (see Figure 6.23). The empirical constants used are $k_1 = \ln(2.0)$ and $1.2 < k_2 < 2.0$. The constant k_2 depends on operational process conditions.
- The spray enthalpy flux carried from the particles onto the deposit is derived from the calculations described in Section 5.1.2.

6.3.3   Fundamental equation and coordinate transformation

The simulation code has been developed for two-dimensional calculations assuming circumferential symmetry of the deposit and the underlying (disc-shaped) substrate. The fundamental conservation equation is the transient heat conduction equation including a source term for phase change modelling and latent heat release. It is used in the derived formulation in enthalpy form:

$$\frac{\partial H}{\partial t} = a \left(\frac{\partial^2 h}{\partial z^2} + \frac{\partial h^2}{\partial r^2} + \frac{1}{r} \frac{\partial h}{\partial r} \right), \tag{6.18}$$

including the temperature conductivity a, the specific enthalpy h and the total enthalpy H, defined as:

$$H = c_p \frac{T - T_{s,l}}{L_h} + f_l, \quad h = c_p \frac{T - T_{s,l}}{L_h}, \tag{6.19}$$

with the latent heat of fusion L_h and the liquid fraction f_l. The relation between specific and total enthalpy is given by:

$$\begin{aligned}
h &= H - 1, & \text{for} \quad H \geq 1 & \quad \text{(fluid);} \\
h &= 0, & \text{for} \quad 0 < H < 1 & \quad \text{(solidifying);} \\
h &= H, & \text{for} \quad H \leq 0 & \quad \text{(solid).}
\end{aligned} \tag{6.20}$$

The local phase and solidification state of the material are deduced directly from the total enthalpy in this formulation.

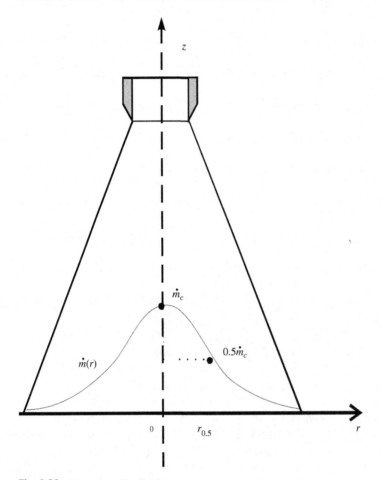

Fig. 6.23 Mass flux distribution

A coordinate transformation of the fundamental equation is performed which transforms the basic equation system into a common curvilinear non-orthogonal coordinate system matching the actual transient deposit contour at each time step. The underlying substrate is assumed to be a round disc that is tackled within a cylindrical coordinate system. The resulting final grid structure (using circumferential and sidewise symmetry, by illustrating a half cut through the deposit and the substrate) is illustrated in Figure 6.24. The coordinate transformation for this special case of a principally Gaussian-shaped deposit uses the height coordinate:

$$z(r) = \frac{\dot{m}_c}{\rho_d} t e^{-k_1 (\frac{r}{r_{0.5}})^{k_2}}. \tag{6.21}$$

As a result, the transformed new coordinates of the system may be derived as:

$$\xi = r, \quad \eta = z e^{\kappa r^{k_2}} \quad \text{with} \quad \kappa = \frac{\ln(2, 0)}{r_{0.5}^{k_2}} \tag{6.22}$$

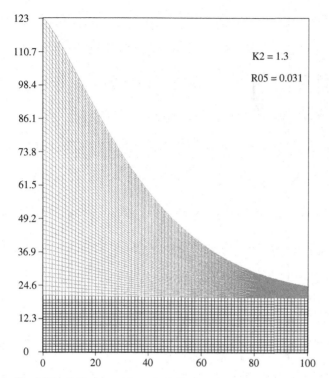

Fig. 6.24   Grid contour for a Gaussian-shaped deposit (Zhang, 1994)

containing the profile constant k_2 of the mass flux distribution in the spray model. In this way, thermal analysis of the deposit is directly coupled to the spray model via the mass flux distribution. Transformation of the conservation equation results in:

$$\frac{1}{a}\frac{\partial H}{\partial t} = Ah_{\xi\xi} + Bh_{\xi} + Ch_{\xi\eta} + Dh_{\eta} + Eh_{\eta\eta}, \tag{6.23}$$

where the spatial derivatives of the local enthalpy h and the geometrical coefficients of the transformation are:

$$A = \xi_r^2 + \xi_z^2,$$
$$B = \xi_{zz}\eta_z + \xi_{rr}\eta_r + \frac{\xi_r}{\overline{\xi}},$$
$$C = 2(\xi_r\eta_r + \xi_z\eta_z), \tag{6.24}$$
$$D = \xi_z\eta_{zz} + \xi_r\eta_{rr} + \frac{\eta_r}{\overline{\xi}},$$
$$E = \eta_r^2 + \eta_z^2.$$

In this series of equations, $\overline{\xi}$ is the radius with respect to the new coordinate system $\overline{\xi} = r(\xi, \eta)$. The mixed derivatives are to be included only for non-orthogonal coordinate transformations.

In order to describe the transient contour of the growing deposit during spray forming, during calculation additional layers need to be added to the top of the geometric model of

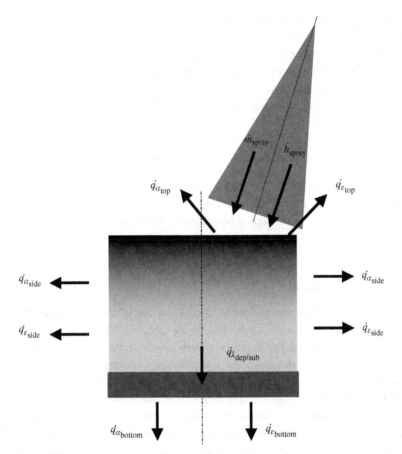

Fig. 6.25 Boundary conditions for thermal billet simulation

the deposit. The layer thickness at each growth step Δt is adapted to the local compacted mass from the spray. Thereby at each growth step, an additional grid layer is added to the system and the total grid size linearly increases in the spray phase.

6.3.4 Boundary conditions

To model the growth of the deposit during thermal simulation, the necessary boundary conditions need to be prescribed. The main processes are:

- heat transfer by convection to the gas phase from the deposit and the substrate,
- heat radiation exchange with the environment,
- local thermal energy and mass flux contributions from the spray entering the deposit surface,
- heat resistance between deposit and substrate from the lower deposit layer (contact layer).

The most important boundary conditions are the fluxes (mass and heat) across the boundary of the solution domain. As illustrated in Figure 6.25 these are:

$\dot{m}_{spray}$ mass flux distribution in the spray,

$\dot{h}_{spray}$ enthalpy flux distribution in the spray,

$\dot{q}_{\alpha_{top}}$ convective heat flux from the top (spray impingement area) of the deposit

$\dot{q}_{\varepsilon_{top}}$ radiative heat flux from the top (spray impingement area) of the deposit

$\dot{q}_{\alpha_{side}}$ convective heat flux from the side of the deposit

$\dot{q}_{\varepsilon_{side}}$ radiative heat flux from the side of the deposit

$\dot{q}_{\lambda_{dep/sub}}$ conductive heat flux from the deposit to the substrate,

$\dot{q}_{\alpha_{bottom}}$ convective heat flux from the bottom of the substrate,

$\dot{q}_{\varepsilon_{bottom}}$ radiative heat flux from the bottom of the substrate.

These boundary conditions are derived either from simulation or from experimental investigation. As an example, in the following, derivation of the surface to gas heat transfer coefficient and the heat resistance coefficient between deposit and substrate will be described.

6.3.5   Heat resistance coefficient

Especially at the start of the spray forming process, cooling of the deposit directly depends on the heat flux from the deposit into the underlying substrate. This heat flux is of the same order as, or may even exceed, the heat flux contribution from the deposit surface to the gas by convection and radiation in some cases. For simulation cooling, analysis and description of this heat flux is a necessary and important boundary condition. The heat flux may not be easily incorporated into the conventional heat conduction model due to resistance effects as will be explained below. A number of resistance effects' need to be accounted for in real spray forming applications, depending on e.g. material pairing (substrate/deposit) and operational process conditions. An experimental method to determine and quantify the heat flux and heat resistance effects between deposit and substrate has been developed by Bergmann (2000), which will be introduced next.

The major influence on heat flow rate $\dot{Q}$ from the deposit to the underlying substrate is the structure and condition of the lowest layer of the deposit (i.e. the first layer to be compacted onto the substrate). This is the contact layer between the sprayed material and the substrate and, due to the process conditions pertaining at the time of formation, may exhibit thermal resistance to heat transfer between the deposit and the substrate. The phenomena affecting thermal resistance may be expressed in terms of surface roughness, porosity, and oxide layer formation. These may result from cold spray conditions and the remaining oxygen in the spray chamber prior to spray forming. A principal sketch of a typical deposit/substrate contact layer is illustrated in Figure 6.26.

The contact layer influences the heat flow rate $\dot{Q}$ by decreasing conductivity in the contact layer compared to that in the deposit under normal conditions. Therefore, for example, measurements of local temperatures just above and below the contact layer show a certain temperature difference in this area (Lavernia and Wu, 1996; Zhang, 1994). The qualitative distribution of the temperature in the contact layer between the deposit and the substrate is shown in Figure 6.27.

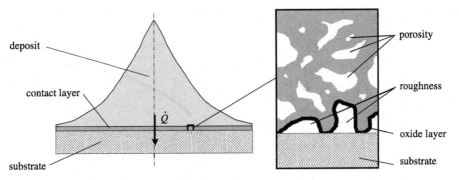

Fig. 6.26 Composition of the contact layer between deposit and substrate

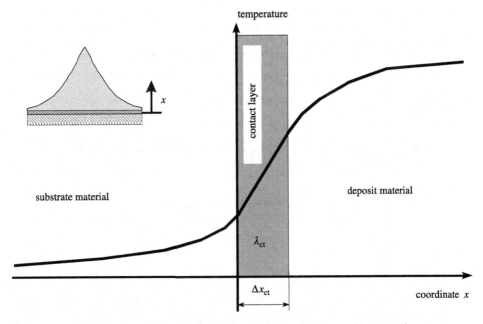

Fig. 6.27 Qualitative temperature distribution in substrate and deposit with contact layer

The heat conductivity of the contact layer is not constant across its thickness Δx_{ct} and depends on the relation between the amount of oxides, the roughness and the local porosity. The assumed heat conductivity λ_{ct} illustrated in Figure 6.27 is an averaged quantity.

Because the contact layer (eventually with a thickness of some millimetres) only contains a minor amount of the total deposit mass and its spatial extension is small, for calculation of the temperature distribution in the deposit detailed resolution and analysis of the composition of the contact layer is of minor importance. Also grid resolution at the scale of the contact layer depth is not attempted during simulation. But the integral influence of the contact layer on the heat transfer mechanism between deposit and substrate needs to be investigated. Here, in analogy to the convective heat transfer mechanism from the surface, a heat

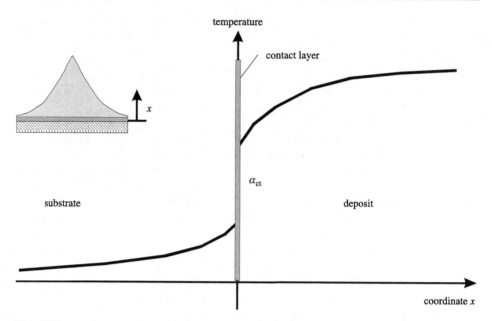

Fig. 6.28 Qualitative temperature distribution in substrate and deposit with heat contact coefficient α_{ct}

contact or resistance coefficient α_{ct} is defined (Baehr and Stephan, 1994). From this point of view, the temperature distribution in the substrate and the deposit may be expressed as in Figure 6.28.

The relation between the average heat conductivity of the contact layer λ_{ct} and the heat resistance coefficient of the contact layer α_{ct} is given by the thickness of the contact layer:

$$\alpha_{ct} = \frac{\lambda_{ct}}{\Delta x_{ct}}. \tag{6.25}$$

For experimental evaluation of the heat resistance coefficient of the contact layer, ex-post model analyses from spray formed products have been performed. Here cylindrical-shaped probes have been taken from the bottom of spray formed preforms and a heat flux applied across the contact layer surface. The stationary temperature distribution in the cylindrical element has been measured. From the assumption of an infinitely thin contact layer and with the known heat flow rate $\dot{Q}_{ext}$, the measured surface temperature at the surface of the heating element and diameter d of the probe, the heat transfer resistance coefficient is:

$$\alpha_{ct} = \frac{\dot{Q}_{ext}}{\frac{\pi}{4} d^2 (T_a - T_b)}. \tag{6.26}$$

The spray forming process conditions and materials used in this study are listed in Table 6.2. The base case for evaluation of the heat resistance coefficient is the one in the first column (standard trial). The other parameter sets deviate in one or more process conditions from this standard parameter set.

Table 6.2 *Model parameters for derivation of the heat contact coefficient*

	Standard	V1	V2	V3
Material	C30	**Cu**	C30	C30
Melt mass flow rate $\dot{M}_l$ [kg/s]	0.192	0.217	0.192	0.192
Gas mass flow rate $\dot{M}_g$ [kg/s]	0.291	0.205	**0.205**	0.291
GMR [-]	1.5	0.9	**1.1**	1.5
Melt superheat ΔT [K]	95	**200**	95	95
Substrate distance z [mm]	750	750	750	750
Substrate material	Steel	Steel	Steel	**Ceramic**

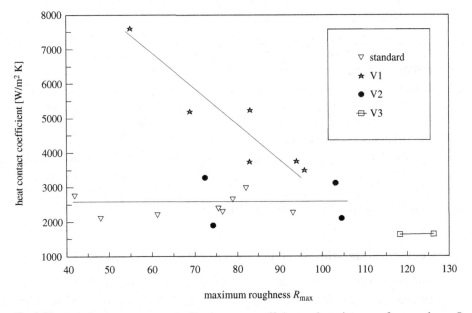

Fig. 6.29 Relation between measured heat contact coefficient and maximum surface roughness R_{max} (for process condition see Table 6.2)

Figure 6.29 illustrates the measured values of the heat resistance coefficient dependent on the measured roughness of the contact layer surface. The individual datum points show samples from different positions along the contact area.

While the samples from deposit V1 (copper) show a pronounced relation between the heat resistance coefficient and the surface roughness of the contact layer (expressed here by means of the measured local maximum roughness R_{max}), within steel spray formed products this behaviour could not be found. For steel spray formed products (standard, V2 and V3), the heat resistance coefficient is almost constant and smaller than for copper. Therefore, the heat flux resistance for steel is lower.

The condition of the contact layer may be directly related to the spray and substrate conditions during impact of the first droplets during spray forming onto the substrate. This is because melt particles that have been cooled more intensively in the spray in the initial phase, impinge onto the (cold) substrate and therefore contain far less melt liquid. Thus, remelting and sintering between particles is hindered, and porosity and roughness in the contact layer are increased. As a result, the resistance coefficient increases. This may explain why within trial V1, for copper, sprayed at higher relative superheating of the melt and at a lower GMR, the measured heat resistance coefficients are far higher compared to the standard case. Also in trial V2, where a smaller GMR when compared to the standard case has been used, the maximum measured value of the heat resistance coefficient also increases.

When using a ceramic substrate material (as in case V3), the heat coefficient is expected to increase due to the lower porosity in the contact layer, which results from less cooling of the droplet layers initially deposited (here the heat flux into the substrate decreases). This effect is not seen, because the ceramic substrates themselves have surface roughness values that are above those measured for the steel substrates in the other cases. This roughness is reflected at the lower surface of the contact layer of the preform during deposition. From this analysis, optimization potentials for spray forming processes may be derived. Coating of the substrate may be an efficient way of controlling the roughness. Due to the relatively high roughness values of the deposits used, the measured values of the contact heat coefficient are higher than expected from the model outlined.

The values of the measured heat contact coefficients are well within the range found in literature. In Lavernia and Wu (1996), for a number of investigations of deposit/substrate heat transfer processes, the range of values for the heat contact coefficient within spray formed products is between 10^3 and 10^4 W/m^2 K.

6.3.6   Convective heat transfer across the surface of a spray formed billet

The distribution of the convective heat transfer coefficient on the surface of a spray formed billet has been numerically investigated by Rau (2002). The simulation model is based on a finite volume representation of the fundamental transport equations for mass and momentum (Eq. (3.1)). A commercial CFD flow solver (FLUENT (creare.x Inc, 1990)) has been used. The main parameters of the simulation are:

- Turbulence modelling is based on a RNG–k–ε model, as this modification of the standard k–ε model has been proved to obtain better results in some flow configurations, including strong streamline curvature cases such as the impinging jet.
- The model is based on a single-phase approach, studying the pure (incompressible) flow of the atomizer gas.
- A fully three-dimensional case is studied: the grid is based on boundary fitting by unstructured grid elements.
- The flow around a billet of dimensions ϕ 200 mm and 300 mm height (typically derived from experiment), where the spray is inclined by 30° to the billet, is studied.

velocity [m/s]

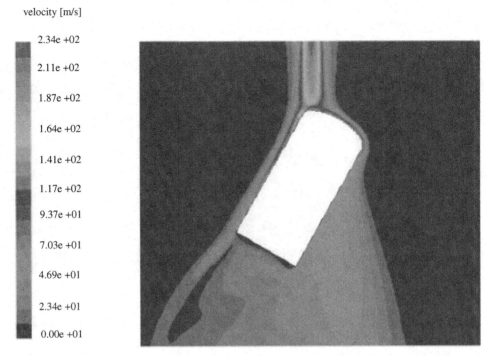

2.34e +02

2.11e +02

1.87e +02

1.64e +02

1.41e +02

1.17e +02

9.37e +01

7.03e +01

4.69e +01

2.34e +01

0.00e +01

Fig. 6.30 Atomizer gas flow field around a billet in the three-dimensional calculation (Rau, 2002)

- The momentum boundary conditions on the billet surface are derived from the 'two-layer zonal model' (creare.x Inc, 1990). This model obtains better resolution of the boundary layer down to the viscous zone.
- Usage of the 'two-layer zonal model' requires a final grid resolution, where the nearest grid point to the wall needs to be located at a maximum dimensionless grid distance of $y^+ < 4 \ldots 5$. The grid distribution in the near-surface area is adapted throughout the iterational procedure and refined in regions where necessary. However, the maximum distance which can be achieved within a domain of times up to 1.2×10^6 grid points is larger than the value mentioned above. Validation of this approach is based on comparison of results for the case of an impinging jet on a flat surface in two-dimensions and the calculated heat transfer coefficient distribution to values published.
- The surface temperature of the billet is assumed to be constant.

The result of the computation for the case of 0.4 MPa atomizer pressure with nitrogen is seen in Figure 6.30. Here the velocity contours in a central plane around the billet are illustrated. Five main flow regions may be distinguished. The jet flow above the billet is characterized by typical free-jet behaviour, such as exponential decrease of the centre-line velocity and the entrainment of ambient gas. On the top of the billet, in the impingement area, a boundary layer resulting from diversion of the jet flow occurs. On the side of the billet, tangentially orientated to the impinging jet, a shearing boundary layer is established.

heat transfer
coefficient [W/m² K]

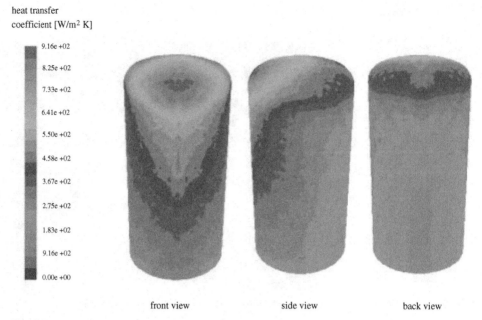

9.16e +02		
8.25e +02		
7.33e +02		
6.41e +02		
5.50e +02		
4.58e +02		
3.67e +02		
2.75e +02		
1.83e +02		
9.16e +02		
0.00e +00		

front view   side view   back view

Fig. 6.31 Isocontour planes of heat transfer coefficient for a billet in an atomizer gas flow (Rau, 2002)

On the shadow side (beneath) of the billet, the flow may detach, resulting in recirculation within that particular area. On the lower side of the billet, recirculation of the gas is also seen.

The distribution of the heat transfer coefficient on the surface of the billet is illustrated as isocontours in Figure 6.31 for 0.4 MPa gas pressure. Maximum values of the heat transfer coefficient in the jet impingement area are about 900 W/m² K; the lowest values are to be found in the shadow region of the billet.

Comparison of the calculated heat transfer coefficient with experimental values is seen in Figure 6.32. The heat transfer coefficient is plotted along a line on the billet surface, as indicated on the right-hand side of Figure 6.32, from the bottom to the top of the billet on its front, across the top of the billet, and down from the top to the bottom on its shadow side. For the experimental investigation of heat transfer coefficients on spray formed billet surfaces, the lumb-capacity method has been used (Schneider *et al.*, 2001; Tillwick, 2000). Here a dummy reflecting the real geometry and size of a spray formed billet has been preheated to approximately 150 °C. The billet is then located beneath the atomizer nozzle inside a spray chamber and is cooled down by the atomizer gas for approximately 60 s. At several points on the billet surface, heat transfer probes are distributed, where the local heat transfer coefficient is measured. Comparison of experimental and simulated results shows agreement of the behaviour of the heat transfer coefficient. But the experimental value in the stagnation point, for example, is measured at just 750 W/m² K. Here maximum grid resolution and the y^+ values of the first grid point are, in some areas of the simulation, much too coarse to resolve the flow structure properly.

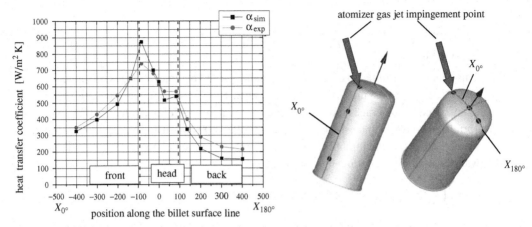

Fig. 6.32 Comparison of simulated (Rau, 2002) and experimental (Tillwick, 2000) heat transfer coefficient distributions along a line (see right side) on the billet

The results shown here describe the local heat transfer coefficient across a stagnant billet surface. However, during a spray forming experiment, the billet rotates around its vertical axis. As temporal resolution of simulation models that calculate the temperature distribution within a spray formed billet in the spray phase is typically much lower than the time increment of the billet rotation, spatial averaging of the heat transfer coefficient on the billet surface is necessary (Schneider *et al.*, 2001; Tillwick, 2000). First, a correlation for the averaged heat transfer coefficient at a constant distance l at the billet top was developed. To calculate the averaged value $\bar{\alpha}$ the local heat transfer coefficient at a constant distance l from the centre of the billet is:

$$\bar{\alpha}(l) = \frac{\sum_{i=1}^{n} \alpha_i}{n}, \tag{6.27}$$

where i stands for different rotation angles φ. In this case, $\Delta\varphi = 90°$ and $\varphi_1 = 0°$. At the centre ($l = 0$ mm) the average value is constant by definition, because the location does not change with rotation angle. The local values just oscillate around the average value at the other locations ($l \neq 0$ mm) at the top of the billet. In the end, one only needs to measure a single position ($l = 0$ mm) to know the averaged value at all locations at the top. This means that the average heat transfer coefficient at the top of the billet only depends on the distance z and the gas mass flow of the nozzle $\dot{M}_G$. In principle, the convective heat transfer of a body in a gas stream depends on the gas velocity and its degree of turbulence. Results of hot wire anemometry measurements in flowing gas around a billet have shown that the turbulence intensity Tu does not change significantly for the parameter range investigated. Therefore, just the maximum velocity u_m at the centre of the jet and the jet velocity half-radius $r_{0.5}$ describe the flow field of the jet stream and the average heat transfer coefficient $\bar{\alpha}$ at the top of the billet ($100 < l < -100$ mm) can be expressed by the following empirical correlation:

$$\bar{\alpha}_{top} \left(\frac{W}{m^2\,K} \right) = 16.2 \cdot u_m \left(\frac{m}{s} \right) \cdot \sqrt{r_{0.5}\ (m)} + 165. \tag{6.28}$$

This equation is limited to the boundary conditions and parameter ranges of Schneider *et al.* (2001): $z = 0.4$ to 0.5 m, $\dot{M}_G = 0.2$ to 0.29 kg/s, Tu $\sim 23\%$, nitrogen and a $30°$ spray angle to the billet. The average heat transfer coefficient at the side of the billet, for one revolution, correlated with the dimensionless height h/H ($H = 400$ mm) using an exponential function, where $h/H = 0$ is the top of the billet, and $h/H = 1$ is the bottom of the billet. The best fit was obtained for the following correlation:

$$\bar{\alpha}_{jacket} \left(\frac{h}{H} \right) = \bar{\alpha}_{top}\ e^{[0.85 \cdot (\frac{h}{H})^2 - 1.65 \cdot (\frac{h}{H})]}. \tag{6.29}$$

Further-refined results from measurements or simulations are necessary to derive more general correlations for the local heat transfer behaviour in spray forming, e.g. in the form of a general Nusselt–Reynolds correlation.

Some general approaches for the overall convective heat transfer coefficient in spray forming have been suggested. Mathur *et al.* (1989) derived a correlation of the Nusselt (Nu) number for convective heat transfer of a preform with respect to the velocity and thermal properties of the impinging gas:

$$Nu = Pr^{0.42} \left(\frac{\delta}{R} \right) \left[\left(1 - 1.1\frac{\delta}{R} \right) I \left(1 + 1.1\frac{z}{\delta - 6}\frac{\delta}{R} \right) \right] 2Re^{0.5}$$
$$\times (1 + Re^{0.55}/200)^{0.5}. \tag{6.30}$$

Here δ is the width of the gas jets at the nozzles, R is the radial distance from the centre-line of the spray, z is the atomizer to substrate distance, Re is the gas Reynolds number and Pr the Prandtl number.

In another approach, Liang and Lavernia (1994) suggested a relationship for the Nusselt number as:

$$Nu = 1.2Re^{0.58} \left(\frac{z}{d_0} \right)^{-0.62}, \tag{6.31}$$

where z is the distance from the gas noozles to the surface and d_0 is the diameter of the nozzle.

6.3.7   Calculated model results for Gaussian deposit cooling

In the following, some thermal simulation results of the deposit/substrate will be discussed, where in the first sequence the heat contact coefficient for the deposit/substrate has not been taken into account. In Figure 6.33 the growth behaviour of a Gaussian-shaped deposit sprayed on a substrate 20 cm in diameter, in a spray sequence of 30 s duration, is shown, together with the calculated temperature distribution. The spray formed material is steel. After 10 s spray time, the heat flux entering the substrate from the deposit has already reached the lower side of the substrate. Therefore, for longer times the deposit does not act

Table 6.3 *Variation of parameters used in the thermal calculation of deposit cooling*

Substrate height h [m]	0.02/0.05
Heat conductivity substrate λ [W/m K]	30/60
Mean liquid content in spray f_l [%]	0/50/100

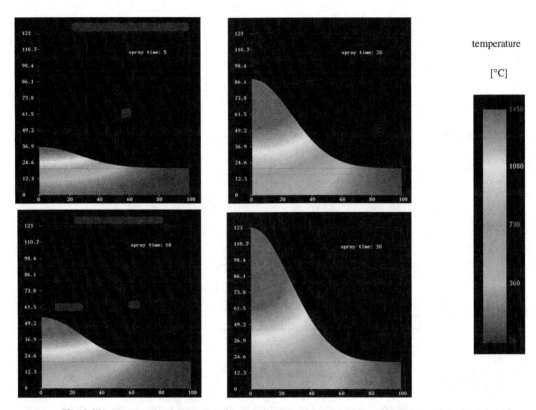

Fig. 6.33 Temperature distribution in the growing deposit and the underlying substrate: spray time measured in seconds (Zhang, 1994)

as a heat sink for the sprayed deposit, storing the thermal energy; now the lower side of the substrate takes place in the overall heat balance by convective cooling.

Four points in the substrate and in the boundary area between substrate and deposit have been chosen for quantitative companion of the simulation results with experimental values. The position of the points is indicated in Figure 6.34. These points have been selected for investigation of the heat contact resistance mechanisms because direct comparison to experimental values has been achieved. Variation of the parameters in the calculations based on the list in Table 6.3 has been done.

Results for simulations based on these parametric variations are illustrated in Figures 6.35 to 6.40. In a first approach, two temporal stages need to be distinguished. Within the initial

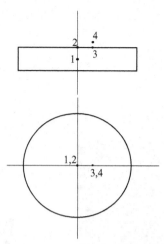

Fig. 6.34 Evaluated positions within deposit and substrate

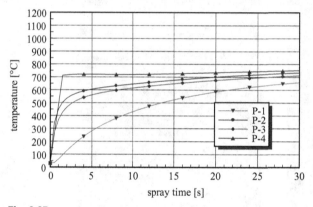

Fig. 6.35 Temporal temperature distribution at points 1 to 4, $f_l = 0\%$ (Zhang, 1994)

three seconds of spray time the temperature at the derived points in the deposit/substrate area increases rapidly, while with increasing spray time the temperature increases far more slowly. The total spray time in the process is assumed to be 30 s in this case.

When the amount of liquid in the impinging spray is increased, as seen in Figures 6.35 to 6.37 for 0 to 100% liquid, the temperatures at the points selected raise continuously. Also the gradients and the increase in temperature at the beginning of the spray phase increase. The calculated temperature difference between points P2 and P3 remains almost constant for different values of the spray liquid contents calculated.

In the spray forming process, the materials of the sprayed particles and that of the substrate may not be identical, introducing an additional degree of freedom in process design. The cooling of the lower deposit layers, for instance, for control of porosity in the bottom of the deposit, may be influenced by choosing the material of the substrate, as well as by introducing additional cooling or heating facilities within the substrate. By accounting for

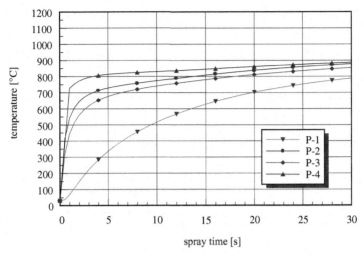

Fig. 6.36 Temporal temperature distribution at points 1 to 4, $f_l = 50\%$ (Zhang, 1994)

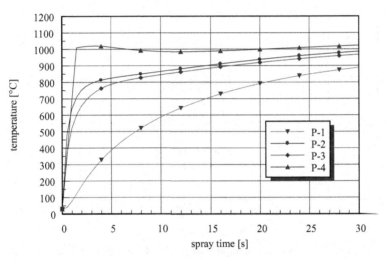

Fig. 6.37 Temporal temperature distribution at points 1 to 4, $f_l = 100\%$ (Zhang, 1994)

the decreased heat conductivity of the substrate in Figure 6.38, heat flow into the substrate is decreased and, as a result, the temperature increase at the centre of the substrate P1 is smaller in comparison. Due to the same reason, the temperature increase in P4 is more pronounced. From this behaviour one can see that small changes and variations of the physical properties of the substrate (as shown here for heat conductivity), may influence the temperature distribution within the deposit. With variation of substrate height (thickness), as shown in Figure 6.39, the temperature of the substrate hardly increases during the initial spray phase (because heat flow from the deposit has not reached the bottom of the substrate).

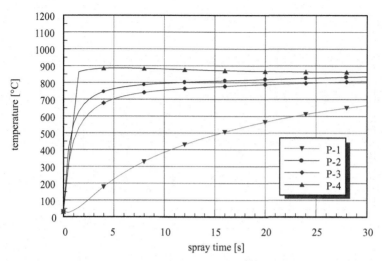

Fig. 6.38   Temporal temperature distribution at points 1 to 4, $\lambda_s = 30$ W/m K (Zhang, 1994)

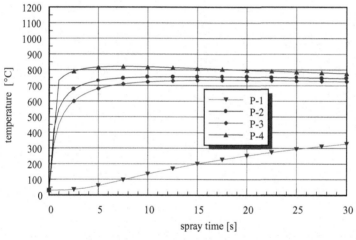

Fig. 6.39   Temporal temperature distribution at points 1 to 4 for a thick substrate ($h = 50$ mm) with $f_l = 50\%$ (Zhang, 1994)

Here the dimension of the substrate may be referred as semi-infinite. At larger spray times, the thicker substrate results in somewhat decreased temperatures at the observed points.

Another point for discussion is shown in Figure 6.40. Here the position of the solidification line after 30 s spray time is illustrated. Above the solidification line, the material is in a mushy state of phase change and still contains some liquid melt. Below the solidification line, the material is fully solidified. Obviously, alteration of the height of the substrate (to 50 mm) only results in small changes in the position of the solidification line at the end of the spray process. But the thermal state of the impinging spray, in terms of liquid content

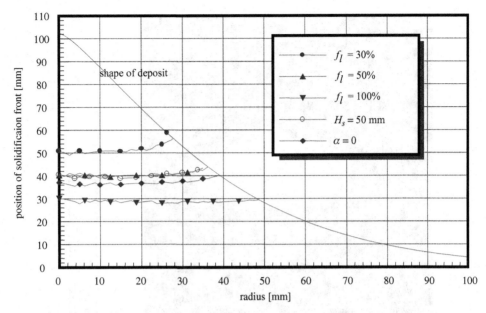

Fig. 6.40 Position of the solidification front at the end of the spray phase ($t = 30$ s) (Zhang, 1994)

of the spray, has a severe influence on the already solidified volume at the end of the spray phase.

The effect of thermal contact resistance between the deposit and the substrate will be discussed next. Here a temperature jump in the contact layer results that influences temperature evaluation and distribution in the deposit. Especially for spray forming processes aimed at the product of large preforms of high volume (such as billets), additional conductive cooling from the substrate is important. Figure 6.41 shows the developing contour of a growing Gaussian preform, together with the surface temperature of the deposit in the spray phase and in the subsequent cooling phase. In Figure 6.41(*a*), the contour and surface temperature after 15 s spray time is illustrated.

The deposit contour and surface temperature at the end of the spray phase, i.e. after 30 s, are illustrated in Figure 6.41(*b*). At this time, spraying ends and the subsequent cooling period starts. The deposit contour does not change further, as can be seen in Figure 6.41(*c*). The surface temperature at this time is lower, as the deposit is cooled by the gas and by the heat flux into the substrate.

A two-dimensional cut through the deposit and the underlying substrate for the same process conditions and times as in the previous figure is illustrated in Figure 6.42. Here, a temperature jump at the deposit/substrate interface due to thermal contact resistance is seen.

By evaluation of the calculated temperature distributions, identification of areas within the deposit is possible where, for longer spray times, high temperatures may occur. Such a hot spot, as seen in Figure 6.42(*c*), may result from the rapidly cooled surface of the

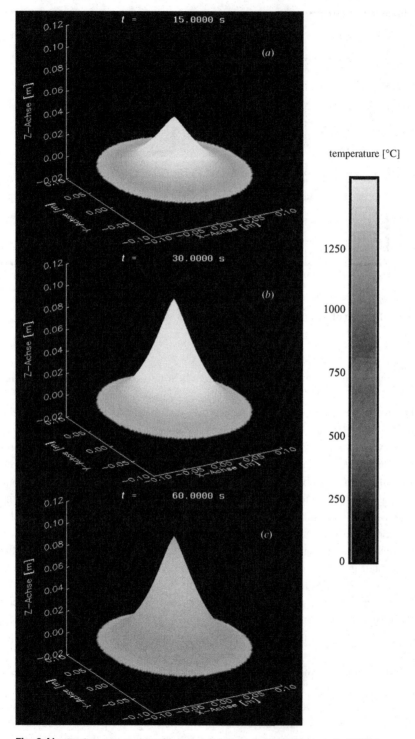

Fig. 6.41   Surface temperature for a growing deposit (Fritsching *et al.*, 1997b)

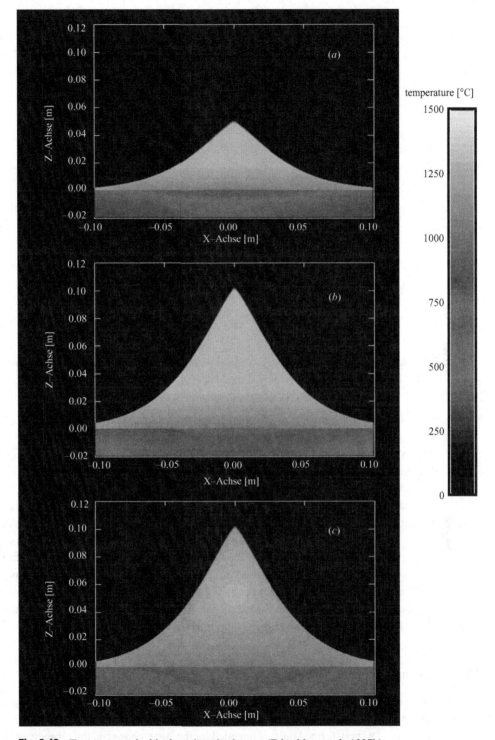

Fig. 6.42 Temperatures inside deposit and substrate (Fritsching *et al.*, 1997b)

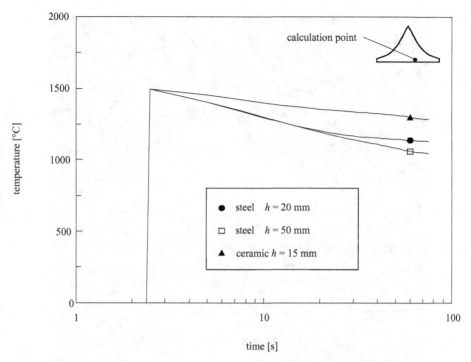

Fig. 6.43   Temperature distributions at a central point inside the deposit, taking into account the heat contact coefficient. Material: steel C30 (Fritsching *et al.*, 1997b)

deposit covering a hotter area beneath. These regions are extremely undesirable in spray forming applications, as grain growth may occur here. In particular, problems may arise if the unsolidified hot spot occurs in a solidified solid shell. Here the difference in material densities may cause high yield stresses and even crack formation in the spray formed product.

Figure 6.43 shows the simulated temperature distribution at a certain point inside the deposit (see sketch). Here the type of substrate material has been changed, altering calculated boundary conditions, while maintaining constant spray process properties and conditions. From this figure it can be seen that the deposit cools most rapidly on the 50 mm steel substrate, while cooling of the deposit onto the ceramic substrate takes longest. A remarkable temperature difference with substrate thickness is not observed in the initial spray phase, but after 30 s spray time, i.e. at the end of the spray phase, and from that point on during cooling, a significant variation in temperature is observed.

6.3.8   Thermal simulation of billet geometry

For thermal simulation of a growing spray-deposited billet, a model and simulation program has been developed by Meyer *et al.* (2000). Here, especially, realistic representation of the geometry of a spray formed billet is intended for the spray phase and the subsequent cooling

Table 6.4 *Standard conditions for the simulation of a copper billet*

parameter	spray phase	cooling phase
Convective heat transfer coefficient billet surface α [W/m² K]	170	10
Mean liquid content in compacting material f_l [kg/kg]	0.5	—
Mean temperature of compacting material T [°C]	1028	—
Temperature of gas in spray chamber T_g [°C]	250	250
Emissivity billet surface ε [-]	0.5	0.5
Convective heat transfer coefficient at bottom of substrate α_s [W/m² K]	450	450
Initial substrate temperature T_s [°C]	30	—
Heat contact coefficient billet/substrate α_{ct} [W/m² K]	1000	1000
Time increment between adding of layer Δt [s]	4.67	—

phase. The program enables the flexible inclusion of different billet-shaped geometries. Calculation of the thermal behaviour of the billet is based on the (aforementioned) solution of the transient heat conduction problem including phase change within the billet contour. The formulation of the energy equation is single phased (fluid and solid are described by a single equation). Modelling of the solidification process is based on an appropriate source term (Voller *et al.*, 1990, 1991). The spray impinging onto the deposit surface consists of solid, semi-solid and liquid particles. During the impingement process, the top layer of the deposit will be created, which is a mixture of fluid and solid material (i.e. a mixing zone or mushy layer).

For the simulation results presented here, the grid system shown on the left-hand side of Figure 6.29 is used. Realization of the grid is based on the outcome of the geometry model introduced in Section 6.2, and aims to reflect a real billet. Therefore, direct comparison between simulated results and actual temperature measurements, which have been performed during a specific spray run, is possible. The conditions of the spray forming process are listed in Table 6.4. The calculations and the experiment are performed for CuSn6.

Until the end of the spray period (which is 360 s in this process run) the height of the billet is steadily increasing. The calculated distribution of the isotherms within the billet is illustrated at a time within the spray phase and at a time within the subsequent cooling phase in Figure 6.44. The temperature level is quite high, and is to be related to the assumed boundary conditions. Here the liquid content of the impinging spray is taken as 50%. After the end of the spray phase, the billet cools down and the liquid material remaining in the billet also solidifies. As the copper material used in this spray run has comparably high heat conductivity, the temperature gradients in the billet, from the top to the bottom of the billet, as well as in radial direction from the centre to the outer edge of the billet, are not very well pronounced. The highest temperature differences are located near the bottom of the billet in the initial phase of the spray run. In the upper part of the billet the temperature gradients are lowest until the end of the spray phase is approached. This may be due to the remaining latent heat within the billet head in the mushy zone at the top of the billet.

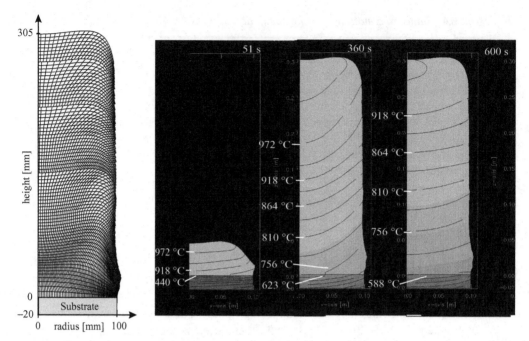

Fig. 6.44 Thermal simulation of a growing billet (Meyer *et al.*, 2000). Left: grid structure. Right: temperature distribution at certain time levels

Comparison of the simulation results based on the chosen boundary conditions shows good agreement with experimental results, as can be seen in Figure 6.45. To allow a direct comparison, the evaluation positions are identical. At the end of the spraying time, the simulation results indicate a more rapid temperature decrease at the bottom of the billet than shown by the measurements. This difference indicates that in the experimental spray run, after the spray phase, less heat has been transferred across the billet surface than indicated by the simulation results. Here the chosen boundary conditions must be checked again and more closely adapted to the experimental process conditions.

One can divide into two categories those boundary and initial conditions that directly influence the thermal distribution in the billet:

- those boundary conditions that can be freely chosen, and
- those coupled parameters that may only be changed in conjunction with other parameters.

Within the first group, the most important parameters are: the entrained heat flux from the spray into the billet and the heat transfer across the billet surface to the gas by convection. These parameters are directly coupled (within the spray phase). Both heat flow rates are directly determined by the atomization parameters of the process. Here, for example, the local distribution of the mass fluxes in the spray, as well as the specific energy content of the spray particles, depend on the GMR (gas to metal flow rate ratio) and other atomization parameters. In the same way, heat transfer to the gas (and therefore the gas temperature in

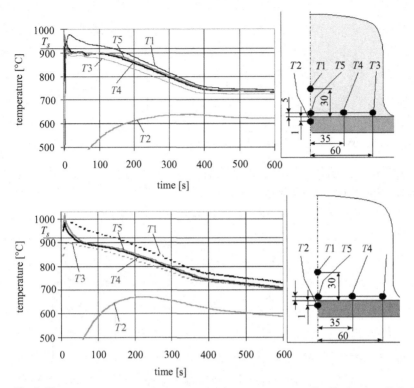

Fig. 6.45 Measured (top) and calculated (bottom) temperature distribution in a CuSn6 billet (Meyer *et al.*, 2000)

the process) and the mean gas velocity around the billet are changed. This couples directly with the heat transfer rate from the billet surface in the spray phase. The same is valid for the heat contact coefficient between billet and deposit. This parameter is mainly influenced by the surface condition of the substrate, the thermal conditions in the spray and in the substrate and, in addition, by the material used in the process.

When the fluid content of the spray is increased, the simulation results in Figure 6.46 indicate a strong increase in the amount of liquid remaining in the billet. Until the end of the spray phase, at $t = 360$ s, the liquid mass in the billet increases steadily. From this simulation it is to be seen that, for the chosen boundary conditions of the spray, steady-state thermal conditions within the growing billet during the spray phase are not achieved, though the geometric shape grows steadily. The thermal conditions in the top of the billet changes throughout the spray process. This may result in changing product quality or material conditions depending on the height of the billet. At the end of the spray phase, the rest of the liquid in the billet solidifies. The time to complete solidification depends on the remaining quantity of liquid melt in the billet. For an assumed fluid content of 10% within the spray, the melt impinging on the billet immediately solidifies at a time increment lower than the resolution time scale of the growing grid system; the mushy zone is less thick

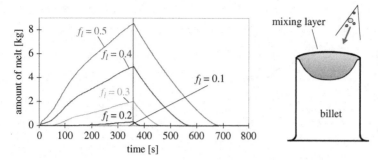

Fig. 6.46   Remaining melt content of the mixing layer (mushy zone), dependent on the liquid content of the spray (Meyer *et al.*, 2000)

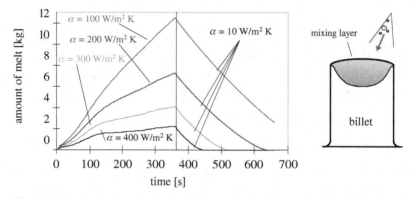

Fig. 6.47   Mass of remaining liquid melt in the mixing layer of a CuSn6 billet for different values of the heat transfer coefficient (Meyer *et al.*, 2000)

than the grid layer. Therefore, each top layer is completely solidified before a new layer of material is added in the simulation. The calculated liquid mass remaining in the billet is zero at all times.

The influence of the heat transfer coefficient on the remaining liquid mass in the billet is seen in Figure 6.47. Assuming different values for the heat transfer coefficient from the billet to the gas within the spray phase, an increasing value of the heat transfer coefficient results in a decreased amount of melt in the billet. By reducing the heat transfer coefficient, less heat is transferred into the gas and the total thermal energy content of the billet is raised. The simulation results indicate the range within which the gas heat transfer coefficient may be varied in order to control the amount of liquid mass in the billet and the thickness of the mushy zone. For an assumed heat transfer coefficient of 400 W/m² K (standard parameter) during the spray phase, the change in liquid mass in the billet is smaller than for a lower heat transfer coefficient of 100 W/m² K. At the beginning of the spray phase, the liquid content in the billet increases rapidly, and after approximately 120 s the increase in liquid mass is somewhat slower. The change in thermal conditions, which have been achieved by means of the changing heat transfer coefficient, may be achieved in a similar way by altering the gas temperature of the environment. As the simulation results indicate, the remaining melt in the billet can be controlled by the operational conditions.

6.3.9   Modification of thermal boundary conditions

Cui *et al.* (2003) have investigated the potential of spray forming as an alternative to conventional casting to minimize possible distortion of steel ball bearing rings during production. In order to do this, homogeneous 100Cr6 steel billets were spray formed with a unique cooling control system. The purpose of the investigation was to control the cooling and solidification behaviour of the deposit more freely. The distribution of the radial thermal profile of the deposit was expected to be more uniform, and a more homogeneous structure achieved. Effects from the following were investigated:

- heating from a furnace around the billet in the spray and cooling phase,
- and/or gas cooling at the bottom of the substrate,
- application of gas flow behaviour inside the furnace on the deposit.

Heat flux modelling was based on the Fourier transport equation and was coupled with the time-dependent geometry of the growing billet, where a perfectly cylindrical-shaped billet has been assumed.

The atomizing gas temperature and the temperature of the spray-chamber wall were assumed to be constant at 250 °C during spray forming and subsequent cooling process. The initial temperature of the low carbon steel substrate was set as 30 °C. In Cui *et al.*'s (2003) model, an average enthalpy method was used, which assumed that the impinging layer had a uniform enthalpy all over the deposit surface. The convective heat transfer coefficient was set as 170 W/m^2 K for standard spray forming conditions. At the end of the spray phase of the melt, the heat transfer coefficient diminished sharply since no additional atomizing gas was sprayed towards the billet. Here the value of 10 W/m^2 K has been taken for calculation for the cooling stage of the billet. The contact heat transfer coefficient (resistance coefficient) between the billet and the substrate was estimated to be 1000 W/m^2 K. The radiative heat transfer emissivity was taken as $\varepsilon = 0.5$ for the computation.

In order to reveal the effects of the heat transfer boundary conditions on the thermal profiles of the spray formed billet, numerical simulation was carried out for side heating, bottom cooling and side isolating (parameters are listed in Table 6.5). In the case of side heating, a furnace was installed around the deposit, and the inside wall temperature of the furnace was set to 1000 °C. For bottom cooling, a gas jet was sprayed onto the bottom surface of substrate to cool it intensively. As for side isolating, the heat transfer coefficient between the deposit and the side of the substrate was reduced to 10 W/m^2 K as a result of protection of the deposit from the atomizing gas flow due to the presence of a shelter (e.g. the furnace around the deposit). In the last case, the combined effect of side heating, bottom cooling and isolation is considered for calculation.

Thermal profiles of the spray formed bearing steel billet were first numerically simulated for the standard spray forming conditions defined in Table 6.5. Figure 6.48 shows the calculated temperature fields and distribution of the residual liquid fraction within the billet

Table 6.5 *Thermal boundary conditions for numerical simulation with thermal variations of billet heating/cooling (Cui et al., 2003)*

boundary condition	temperature of chamber wall, T_{wall} [°C]		heat transfer coefficient at deposit and side of substrate, α_g [W/m² K]		heat transfer coefficient at the bottom surface of substrate, α_{bottom} [W/m² K]	
	deposition	post-deposition	deposition	post-deposition	deposition	post-deposition
(a) standard	250	250	170	10	170	10
(b) side heating	1000	1000	170	10	170	10
(c) bottom cooling	250	250	170	10	500	500
(d) side isolating	250	250	10	10	170	10
(e) (b) + (c) + (d)	1000	1000	10	10	500	500

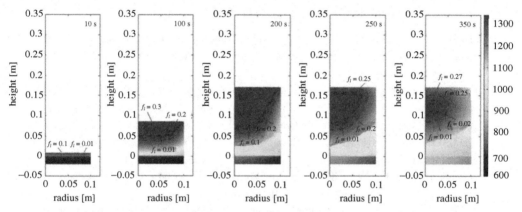

Fig. 6.48 Calculated temperature and distribution of residual liquid fraction in a cylindrical deposit and substrate during spray forming (Cui *et al.*, 2003)

at different spraying times. At the beginning of deposition, the material is almost completely solidified due to the chilling effect of the cold substrate. The overall temperature field and local liquid fraction in the billet increase as the billet grows, i.e. the thickness of the total liquid mass or mushy zone at the top of the billet increases as well. At the end of the spray phase (here $t = 200$ s), the highest residual liquid fraction reaches approximately 0.3. Because of the low thermal conductivity of the 100Cr6 bearing steel, the temperature difference between the billet top and bottom, as well as along the radius, is relatively high. Even when the convective heat transfer coefficient drops sharply at the end of spraying, an uneven radial temperature distribution is still expected.

The local solidification time has been used to demonstrate the cooling and solidification behaviour of the deposit, because the solidification microstructure is largely dependent on this profile. As shown in Figure 6.49, isochronous solidification lines are plotted for the deposit from 10 to 300 s. These lines indicate when and where the solidification process will finish within the deposit. The longer the solidification time, the lower the cooling

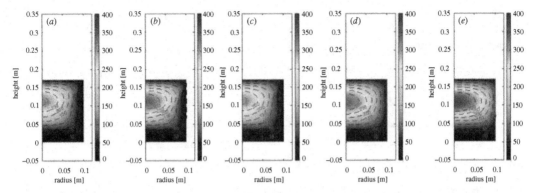

Fig. 6.49 Local solidification time of spray formed 100Cr6 billet under various boundary conditions: (*a*) standard, (*b*) side heating, (*c*) bottom cooling, (*d*) side isolating and (*e*) (*b*) + (*c*) + (*d*) (Cui *et al.*, 2003)

and solidification rates. As mentioned before, porosity at the periphery of a spray formed billet is always a problem, due to rapid cooling of the depositing material in that area. If the local solidification time in the periphery can be prolonged, it would favour elimination of this shrinkage porosity. Figure 6.49(*b*) shows the solidification profile of deposit during side heating, where a heating device at high temperature is placed around the deposit. It can be seen that all isochronous lines move to the right, showing that it takes a longer time for the deposit to solidify. For bottom cooling, see Figure 6.49(*c*), no evident change in the plot can be found compared to the standard case, Figure 6.49(*a*), due to the low conductivity of the steel substrate and the limited thermal influence. As the deposit is isolated from the atomizing gas jet, the convective heat transfer coefficient will decrease markedly, resulting in less heat transfer from the side of the deposit to the gaseous environment. Accordingly, the isochronal lines in Figure 6.49(*d*) move to the right as well. When the effects of side heating, bottom cooling and isolation are combined, a much greater change in the local solidification time of the deposit is noticed, as can be seen in Figure 6.49(*e*). As discovered from detailed calculation of the local solidification time, fast cooling and solidification of the billet at the periphery, with respect to standard processing conditions, can modify the thermal boundary conditions.

The effect of control of cooling on the thermal profiles of the billet is also reflected in the heat flows at the surface of deposit; as shown in Figure 6.50, for different modes and directions. For the standard spray condition, Figure 6.50(*a*), heat flow by convection at the bottom of the deposit is very large because of the chilling effect of the cold substrate at the beginning of deposition. With on-going deposition, the substrate temperature increases rapidly, causing the heat flow to decrease quickly. Flow of heat from the side of the deposit by gas convection and radiation increases nearly linearly as the deposit grows. The heat flow released from the top surface of the deposit maintains a relatively stable value during the spray period because the temperature here normally remains constant due to the continuous spraying of additional droplets. At the end of the spray phase, due to convection at the top and side of the deposit, the flow of heat drops dramatically as atomization has stopped. Radiative heat transfer from the side and top of the billet, as well as heat flow at the deposit/substrate

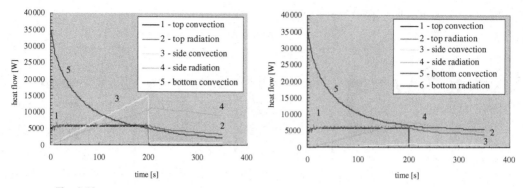

Fig. 6.50   Heat flow at the surface of a cylindrical deposit in various modes and directions in spray and cooling phases (Cui *et al.*, 2003)

interface, also decrease gradually since the temperature of the billet decreases during the subsequent cooling period. Thus, by controlling the cooling conditions, Figure 6.50(*b*), i.e. involving side heating from the furnace, bottom cooling by the gas and reduced gas flow at the side of the deposit, convective heat flow from the side of the deposit and radiative heat transfer are greatly diminished. The flow of heat from the bottom of the deposit to the substrate is elevated owing to enhanced gas cooling of substrate, but this effect is not very significant. No visible change is found in the heat flow from the top surface of deposit because the heat transfer boundary condition has not been modified here.

Computation and analysis have shown that for this case of ball bearing steel spray forming it is quite difficult to obtain an absolutely uniform thermal profile along the radius of the billet because of (1) existing heat flows at the free surface of the deposit and (2) low heat conductivity of the steel. Nevertheless, control of cooling and solidification of the deposit, especially at its peripheries, could be affected to some extent by adjusting the boundary conditions of heat transfer.

Thermal simulation of a growing spray formed billet within so-called 'clean' spray forming process (Carter *et al.*, 1999) has been performed by Minisandram *et al.* (2000). The term 'clean' represents a process where the melt, during heating and delivery, has no contact with ceramic or metallic elements. Therefore, the final product is not contaminated by ceramic or metallic inclusions. This is an important condition especially for superalloy spray forming, e.g. for turbine component (rings) production. Simulation of the overall thermal process for billet production has been divided into several steps/models:

(1) a spray model (axisymmetric, including thermal and momentum coupling);
(2) a geometry model (including an experimentally determined compaction rate);
(3) a thermal billet model, subdivided into two separate spatial regions:
    (3a) a billet surface model,
    (3b) a full billet model.

The surface model describes the thermal conditions only at the surface of the growing billet during the spray phase. The impinging spray is resolved in time and space (scanning

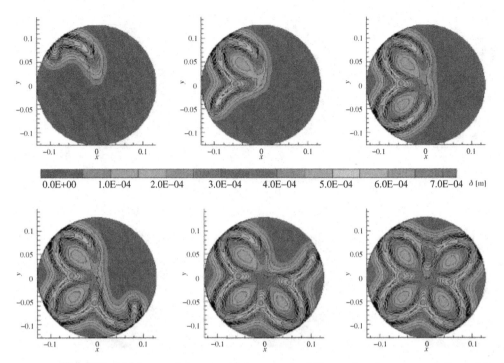

Fig. 6.51 Thickness of the sprayed layer at the billet surface for one revolution of the billet (Minisandram *et al.*, 2000)

mechanism), also rotation of the billet is accounted for. From this model the surface temperature as well as the averaged (mean value, averaged for circumference of the billet surface) coefficients for convective and radiative heat transfer at the billet surface are derived. The main assumption of the surface model is that heat transport in the plane of the billet surface can be neglected compared to transport normal to the surface. Therefore, the heat balance for the deposited layer is derived independently for every part of the surface. The energy balance of the growing layer includes heat losses by convection and radiation as well as by conduction to the underlying layer. Enthalpy is introduced via the spray from above. The base temperature of each surface element is the corresponding temperature value of the billet taken from the full billet model at the end of the preceding time step. Time resolution is sufficient to resolve the actual scanning motion of the atomizer and the rotation frequency of the billet. Evaluation of the surface temperature within the surface model is combined with the axisymmetric model of the whole billet for derivation of the thermal behaviour of the growing billet within the spray phase.

Figures 6.51 and 6.52 illustrate the results of the surface model for the first macro time step for one revolution of the billet. Shown are the variation of billet height (thickness of the sprayed layer), Figure 6.51, and the distribution of the surface temperature, Figure 6.52. At the first time step, the spray impinges onto a plane surface of constant temperature. The end of this first time phase has been chosen here in order to visualize the results clearly. The leaf-shaped structure of the sprayed layer corresponds with the scanning motion of

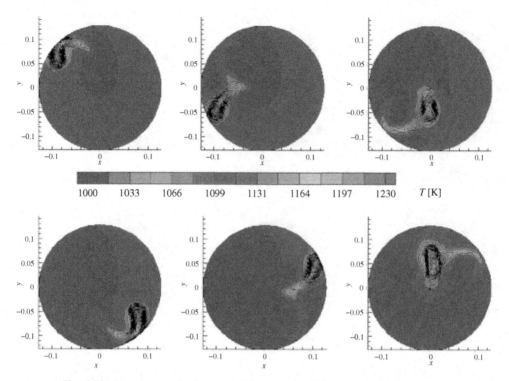

Fig. 6.52 Temperature in the sprayed layer at the billet surface for one revolution of the billet (Minisandram *et al.*, 2000)

the atomizer and the rotational motion of the billet. The temperature distribution illustrates that the temperature in the last sprayed layer is highest and decreases as the spray moves across the billet surface.

6.4   Material properties

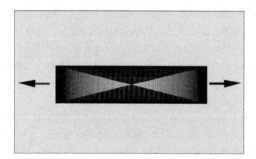

Numerical models and simulations have been introduced into almost all engineering disciplines. Classical materials science and materials technology models may contribute to the development of spray forming processes, predicting a priori from prescribed processes and boundary conditions, the main material properties of spray formed products.

Boundary and initial conditions for material analysis and simulation during spray forming may be derived from the submodels aforementioned for the analysis of impacting particles and the thermal state of the deposit. Thermal conditions and local cooling behaviour mainly influence the material properties, as well as related properties such as grain size, porosity distribution or residual stresses.

6.4.1   Residual stress modelling

Kienzler and Schröder (1997) investigated the characteristic quantities of the material by modelling within spray forming. Input models for their calculations of the residual stress and porosity distribution of spray formed deposits are appropriate material and damage models. The residual stress calculations used are tailored for specific materials or groups of materials. In order to qualify for inclusion in their models, the materials selected must be describable in terms of the following phenomena:

- plasticity and viscoplasticity,
- creep and relaxation behaviour.

In a general material model, some parameters for that specific material used need to be introduced, which may be derived from experimental data. Complete sets of parameters are available only for a limited number of materials. Kienzler and Schröder (1997) have incorporated such material models into a commercial finite element simulation code and used this to calculate the material properties in two-dimensional spray formed Gaussian- and billet-shaped preform deposits.

Figure 6.53 shows the calculated distribution of residual stresses and the distribution of the circumferential stress after spraying and cooling of the material to room temperature. This calculation has been performed for a Gaussian-shaped deposit. The material under investigation is a nickel-based alloy IN738LC. In the core of the deposit, tensile stresses are to be seen; while in the outer regions of the deposit, compressive stresses are found. These stress distributions result from the growth and cooling conditions of the sprayed deposit. The outer area of the deposit cools down faster than the core due to convective and radiative heat transfer from the surface of the deposit. This behaviour results in strain impediments that are reflected in the calculated stress distribution in the deposit.

6.4.2   Macro- and micopore formation

Problems in spray forming may arise from gas pores remaining in the preform. These may occur, for example, due to gas entrapment (too hot a spraying condition) or insufficient liquid content in the spray (too cold a spraying condition). The possible formation of macro- and micropores in spray formed deposits has been investigated during modelling of the spray process of a liquid metal, which solidifies at an arbitrary surface, during the

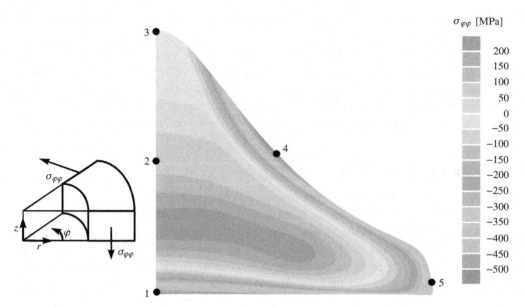

Fig. 6.53 Calculated residual stress distribution (circumferential stress) of a spray formed deposit of Gaussian shape (Kienzler and Schröder, 1997)

geometric simulation of Djuric and coworkers (1999, 2001). Results of their modelling during simulated notch filling have been discussed in Section 6.2.

6.4.3   Porosity model

As pointed out, for example, by Cai and Lavernia (1997), the retention of some porosity is inevitable in spray formed materials. Typically, zero or low porosity (<1%) is found over a large area of the core of spray formed preforms, but the porosity level may increase up to 10% within the surface area of the preform. From a quality point of view, the porosity in spray deposited materials yields two problems. First, material properties may degrade due to the presence of pores. Second, secondary working (such as extrusion, rolling, forging and HIPing) is necessary to achieve full density, which in turn limits the suitability of spray forming as a near-net shape manufacturing process. Consequently, research and modelling efforts have been initiated in spray forming to understand the formation mechanisms of porosity formation better and to determine ways to decrease the porosity level either by adjusting the process parameters or by exploring alternative approaches.

The total porosity in an as-deposited material in spray forming can be described as the sum of:

- porosity from gas entrapment,
- interstitial porosity,
- porosity associated with shrinkage during solidification.

Among these mechanisms, interstitial porosity has been reported as playing the most important role in porosity formation during spray forming, particularly when spraying thin geometries such as bands and strips or tube clads. Considering that the formation of interstitial porosity is due to the lack of liquid to fill overlapping particle interstices, whereas solidification porosity is significant when too much liquid is present in the spray upon impingement, interstitial porosity and solidification porosity may be regarded as mutually exclusive. Consequently, in a modelling approach, when interstitial porosity predominates, solidification porosity is ignored (and vice versa). In a first modelling approach (Cai and Lavernia, 1997), gas entrapment porosity has been neglected. The model describes a two-stage consolidation mechanism. In the first stage, solidified droplets impinge on the deposition surface and form a random dense particle packing. During the second stage, liquid droplets first impinge on the resultant particle packing and, finally, solidify. If the volume of liquid, once solidified, is smaller than that of voids in the particle packing, porosity is assumed to be dominated by interstitial porosity, whereas solidification shrinkage is neglected. On the other hand, if the volume of liquid, once solidified, is more than that corresponding to voids, porosity is assumed to be controlled by solidification of the liquid left and interstitial porosity is neglected – solidification porosity dominates. For this model, a log-normal particle-size distribution has been assumed in the spray, the median particle diameter has been adopted from Lubanska's formula (Eq. (4.49)), and the standard deviation of the drop-size distribution has been taken from empirical metal atomization data (Lawley, 1992).

The porosity model has been applied to spray forming of an Al–4 wt. % Cu alloy, atomized by nitrogen. The main process parameters varied are: the atomization gas pressure, in a range from 1 to 10 MPa; the melt flow rate, between 0.001 and 0.02 kg/s; and melt superheating, from 50 to 250 K. The temperature of the atomization gas is assumed to be constant. A typical result of the simulation is shown in Figure 6.54. Here, the variation of porosity with deposition distance for an atomization gas pressure of 1.2 MPa, a melt flow rate of 0.01 kg/s and varying melt superheating is illustrated. The porosity exhibits a V-shaped variation with deposition distance, reflecting the development of both solidification and interstitial porosity. For small deposition distances, the spray is too hot and solidification porosity results. As deposition distance increases, porosity decreases continuously to the point at which a minimum porosity is achieved. Beyond this point, porosity increases almost linearly with increasing deposition distance. Here, the spray conditions become too cold and the porosity is caused by interstitial effects. As an example, for a melt superheat of 100 K, porosity is 5% at a deposition distance of 0.1 m. The point corresponding to a minimum porosity in this case is at a deposition distance of 0.36 m. For increasing melt superheat, the minimum porosity is achieved at greater deposition distances.

Variation of other process parameters, such as the atomization gas pressure and the melt flow rate, results in principle in V-shaped porosity curves similar to those observed for variable deposition distance. Based on these results, optimum combinations of processing parameters during spray forming may be identified. A porosity coefficient may be introduced (Cai and Lavernia, 1997) to identify dominant mechanisms in the different porosity formation regimes. In addition to porosity, other process constraints, such as microstructure

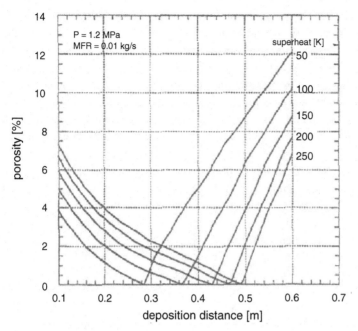

Fig. 6.54 Porosity as a function of deposition distance for an atomization gas pressure of 1.2 MPa and a melt flow rate of 0.01 kg/s for Al–4 wt. % Cu (Cai and Lavernia, 1997)

as well as process economy, have to be taken into consideration to find a set of suitable spray forming process conditions.

Cai and Lavernia's (1997) model of porosity formation during spray forming only takes spray conditions into account. To develop the model further, the influence of the state of the deposit surface on porosity has to be taken into account. This may be achieved by controlling the surface temperature of the deposit, which severely affects the porosity as well as the particle sticking efficiency (Buchholz, 2002).

6.4.4   Microstructure modelling and material properties

One of the main advantages of spray formed materials is the associated equiaxed fine grain sizes and refined microstructure typically to be found. Typical microstructure features of cast materials, such as large columnar grains and the presence of dendrites, may be reduced or eliminated within spray deposition processes. On the other hand, spray formed materials often tend to exhibit greater porosity than found in cast materials. Post-treatment processes (a machining process or heat treatment) are therefore a necessity during spray forming to reduce the porosity level. However, concurrent and additional heat treatment at sufficiently high temperatures may result in undesired grain growth. Therefore, the properties of spray formed materials also need to be described after post-processing as well as in the as-sprayed condition.

A review of microstructural properties of droplet-based manufacturing processes has been given by Armster *et al.* (2002), who found that four primary factors contribute to small grain sizes and grain refinement in solidifying metal materials:

- high cooling rates of the melt,
- high undercooling prior to nucleation,
- the presence or absence of impurities and secondary nuclei in the melt,
- fragmentation of formerly grown dendrites.

To build a certain lattice arrangement during phase change, two conditions must be fulfilled: melt elements need to have a certain amount of time and a certain amount of free energy. For rapidly cooled particles within spray processes, the time for ordering of atoms is quite low and the free path length for single atoms is minimized. Thereby a number of regular-shaped grains or clusters are formed at sizes that are inversely proportional to the element cooling rate. At low cooling rates, the short-range order gives way to preferred growth directions in the melt, resulting in the formation of large grains. Also, increased time for cooling enables simultaneous precipitation of chemical phases and the formation of larger grains. In addition, extended grain growth depends upon the presence of heterogeneous nucleation centres, e.g. from impurities in the melt. Such heterogeneous nuclei may congregate at the grain boundaries, thereby reducing the interfacial energy between grains and retarding grain growth. In spray deposition processes, this means, for example, that the concentration of reactive elements in the processing gas may also be used to control grain refinement and grain growth. Finally, possible fragmentation of dendrites during droplet impact, caused by fluid motion within the droplet, may also contribute to the final formation of small grains.

Investigations concerning the relationship between processing parameters and grain size and refinement in metal spray deposition processes are mainly performed within single-drop experiments or within specific spray deposition arrangements (see, for example, Norman *et al.* (1998); Passow *et al.* (1993); Schmaltz and Amon (1995)). The usefulness of microstructure selection maps and the derivation of empirical correlations between microstructure, droplet undercooling and material composition in industrial droplet-based manufacturing processes has been investigated by Norman *et al.* (1998) employing single-drop experiments, for example, from drop tubes and drop levitation and spray deposition. By comparing the achievable undercooling in well-controlled experiments of relatively large levitated droplets (several millimetres) and correlating the microstructure to spray-process-related drop sizes in the micrometre range, these authors have shown that nucleation is a reasonably well-behaved function of drop diameter. Results for levitated droplets show that typical bands of grain refinement exist at high and low undercooling rates as well. Within the middle range of moderate undercooling, the primary microstructure is coarse grained and dendritic. This result is shown in Figure 6.55 for Ni–Cu droplets at various mixture concentrations. However, the grain morphology is quite different for low and high undercooling rates. For the low undercooling regime, the microstructure consists largely of fragmented dendrites. At the high undercooling regime, larger equiaxed grains are observed. Therefore, two different refinement processes may be distinguished for low and high undercooling rates. For Ni–Cu

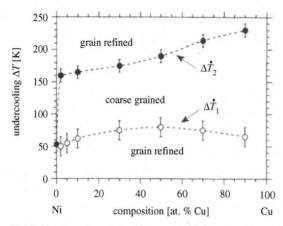

Fig. 6.55 Experimental microstructure selection map for Ni–Cu particles dependent on undercooling: from levitation experiment (Norman *et al.*, 1998)

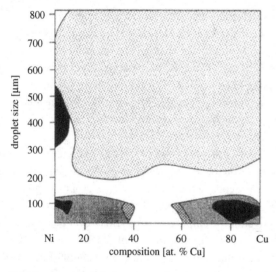

predominantly dendritic
predominantly grain refined at high undercooling rate
predominantly grain refined at low undercooling rate
mixture of grain refined at high and low undercooling rates
mixture of dendritic and grain refined (high and low)

Fig. 6.56 Microstructure predominance map for Ni–Cu droplets produced from drop tube and atomization experiments (Norman *et al.*, 1998)

droplets in the size range up to 800 μm (from drop tube and atomization experiments) refinement is inversely proportional to droplet size, which implies that the amount of undercooling increases for the smallest droplets, as shown in Figure 6.56. The larger droplets (above 600 μm) predominantly show dendritic microstructures. The smaller droplets (below 250 μm) exhibit a grain-refined microstructure and the medium-sized droplets (250

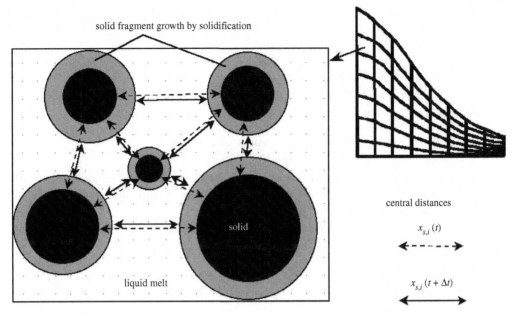

solid fragment growth by solidification

central distances

$x_{s,i}(t)$

solid

liquid melt

$x_{s,i}(t + \Delta t)$

Fig. 6.57 Solidification model inside a volume element in the mixing layer: temporal change of distance between the solidification fronts (two-dimensional representation) (Bergmann, 2000)

to 600 μm) show a transitional combination of dendritic and refined microstructure. Such microstructural predominance maps are available only for a very limited number of alloys, but may be used for incorporation into a coupled model of final microstructure analysis within spray forming.

6.4.5   Grain-size modelling in spray forming

A model for the factors controlling spray formed grain sizes has been presented by Grant (1998) and has been continued, for example, by Bergmann (2002) to analyse grain formation and final grain sizes within spray formed preforms of copper and steel. Because of the dispersed character of the mixing layer at the top of a spray formed specimen during the spray phase, no well-defined solidification front exists in this area. The material solidifies from the surface of a solid particle (or fragment) in all directions (as illustrated in Figure 6.57) until the individual solidification fronts approach each other and merge together. For analysis of this process, a mean distance x_s between solid particles, surrounded by liquid melt, within the mixing layer is defined, dependent on the spray properties and the local fraction of solids. During solidification in the mixing layer, the solidifying material accreted at the surface of the already solid fragments. Based on the following assumptions:

- the solid fragments in the mixing layer are initially almost spherical,
- the initial solid fragments are equally distributed within this zone, and
- the densities of solid and liquid are almost identical,

the mean distance x_s between solid fragments can be described as:

$$x_s = \frac{2}{3} \frac{1 - f_s}{f_s} d'_{3,2}.$$ (6.32)

The derived Sauter mean diameter $d'_{3,2}$ describes the particle-size distribution of the solid fragments immediately after compaction in the top zone of the mixing layer. Following Grant's (1998) proposal, this particle-size distribution is mainly influenced by:

- the original particle-size distribution in the spray,
- the mean local solid fraction of impinging particles $f_{s,0}(r)$,
- the critical impact particle diameter d_{crit} (all particles smaller than this diameter will be deflected by the gas and will not be deposited – they contribute to the aerodynamic overspray (Kramer $et\ al.$, 1997)), and
- the number of solid fragments resulting from each impacting particle (partially or fully solidified particles which may be fragmented during impact).

The resulting solid fragment mean diameter in the mixing layer from an individual particle impacting at the top of the deposit is described (Grant, 1998) as:

$$d_0 = \sqrt[3]{\frac{\overline{f_{s,0}}(r) D(d_p) d_p^3}{G(d_p)}}.$$ (6.33)

The function $D(d_p)$ is a binary switch function, where $D = 0$ if d_p is smaller than the critical impact diameter ($\sim 10\ \mu$m), otherwise $D = 1$. The parameter $G(d_p)$ describes the number of solid fragments resulting from the impact of a d_p-sized particle onto the deposit surface (fragmentation rate). For $G(d_p)$ a functional behaviour is suggested, where for a (spray forming typical mean) particle diameter of $d_p = 100\ \mu$m in the spray, a maximum value of 10 is proposed and decreasing values for bigger and smaller particles. A minimum value of 1 is achieved for $d_p = 50\ \mu$m (and smaller) and $d_p = 200\ \mu$m (and bigger). For simplification of the model it has been assumed that all particles (regardless of size) are compacting and that the particle-size distribution in the spray is not a function of the local (radial) position and corresponds to the totally averaged particle-size distribution. Therefore, $D(d_p) = 1$ and:

$$d_0 = \sqrt[3]{\frac{\overline{f_{s,0}}(r)}{G(d_p)}} d_p.$$ (6.34)

If a constant value for $G(d_p)$ is assumed, the number density distribution of particles in the spray equals that of the solid fragments in the mixing layer $q_0(d_0) = q_0(d_p)$.

Therefore the Sauter mean diameter $d'_{3,2}$ of the solid fragment size distribution in the mixing layer is:

$$d'_{3,2} = \frac{\int d_0^3\, q_0(d_0)\, dd_0}{\int d_0^2\, q_0(d_0)\, dd_0} = \frac{\frac{\overline{f_{s,0}}(r)}{G}}{\left(\sqrt[3]{\frac{\overline{f_{s,0}}(r)}{G}}\right)^2} \frac{\int d_p^3\, q_0(d_p)\, dd_p}{\int d_p^2\, q_0(d_p)\, dd_p} = \sqrt[3]{\frac{\overline{f_{s,0}}(r)}{G}}\, d_{3,2}.$$ (6.35)

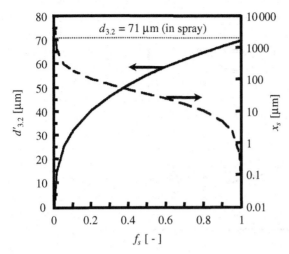

Fig. 6.58 Relation between Sauter diameter of the solid fraction in the mixing layer $d'_{3.2}$ (for $G = 1$: no disintegration), the mean distance between solid particles x_s and the mean local solidification fraction in the mixing layer f_s (Bergmann, 2000)

In this equation, $d_{3.2}$ is the Sauter mean diameter of the original particle size distribution in the spray. Typically, measured particle-size distributions within steel spray forming experiments have yielded a Sauter diameter of, e.g. 63 μm (GMR = 1.5) to 71 μm (GMR = 1.1). The mean distance between solid fragments immediately after deposition in the mixing layer therefore can be described as:

$$x_s = \frac{2}{3} \frac{1 - \overline{f}_{s,0}(r)}{\overline{f}_{s,0}(r)} \sqrt[3]{\frac{\overline{f}_{s,0}(r)}{G}} \, d_{3.2}. \tag{6.36}$$

Based on the following assumptions:

• that the number of solid fragments during the solidification step does not change (neglecting complete remelting of solid fragments and also additional nucleation),
• that the growth velocity is independent of size and only influenced by the local temperature difference, and
• that the mean local solid content of the impinging droplets $f_{s,0}(r)$ can be expressed by the instantaneous local solid content in a volume element of the mixing layer f_s,

the mean instantaneous fragment distance can be described during the whole process.

Results of a grain-size model calculation based on the above assumptions are shown in Figure 6.58 (in a first approach for $G = 1$: assuming no fragmentation). The mean distance between solid fragments x_s continuously decreases with increasing total solid fractions, because the solid particles within the solidification process grow towards each other. Hereby the total area of the solidification front (this is the total surface of all solid fragments) steadily increases. Therefore, the change in mean distance decreases with increasing solid fraction (logarithmic scale) as the new solidified mass is distributed over a larger area. Due to the

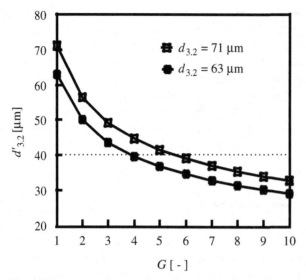

Fig. 6.59 Dependency of the Sauter diameter $d'_{3.2}$ in the mixing layer on the mean disintegration rate G for different particle Sauter mean diameters in the spray $d_{3.2} = 71$ μm and $d_{3.2} = 63$ μm (Bergmann, 2000)

same reason, the Sauter diameter decreases with increasing solid fraction. As seen from Figure 6.58 the Sauter mean diameter for the solid fragments in the mixing layer at the end of the solidification process ($f_s = 1$) reaches the value of the Sauter mean diameter in the spray (here, for example, 71 μm). In this case it is assumed that, during deposition, no fragmentation of solidified droplets/particles occurs.

In publications focusing on grain-size distribution in spray formed products, Grant (1998), Jordan and Harig (1998) and Lavernia and Wu (1996), typical mean grain sizes of about 20 to 40 μm have been found. Because of the homogeneous grain-size distribution typically found in spray formed materials, the actual size of individual grains only slightly deviates from this mean value. Therefore, its grain Sauter mean diameter is in the same size range and it is somewhat smaller than the Sauter mean diameter of the particles in the spray. Thus, for the mean fragmentation function a value greater than 1 has to be chosen. Figure 6.59 shows the dependency of the mixing layer Sauter mean diameter on the mean fragmentation value. A mean fragmentation value $4 < G < 6$ yields mean Sauter diameters in the mixing layer of about 40 μm. Here (on average), from each compacting particle in the spray, five solid fragments in the deposit result. Therefore, a value of $G = 5$ is assumed and the resulting distributions are illustrated in Figure 6.60. Because of adjustment of the fragmentation function to a constant value $G = 5$, the final value of the Sauter mean diameter for the completely solidified material ($f_s = 1$) is somewhat smaller and is in the size range of the experimentally determined grain-size distribution of spray formed deposits. As the Sauter mean diameter is the scaling value for the mean solid fragment distance in the mixing layer, the distribution of x_s for $G = 5$ is below that for $G = 1$.

In summary, modelling of compaction and preform and structure evolution with respect to the integral spray forming process, suggests some possible ways in which process

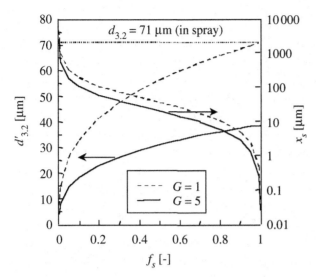

Fig. 6.60 Relation between Sauter diameter $d'_{3.2}$ of the solid fraction in the mixing layer, the mean distance between solid particles x_s and the mean local solidification fraction in the mixing layer for disintegration rates of $G = 1$ and $G = 5$ (Bergmann, 2000)

conditions may be controlled. Investigation of thermal cooling conditions of spray formed deposits, both in the spray phase and in the cooling phase, outlines the boundary conditions affecting material properties of the product. Thus one may start to derive suitable process control mechanisms and quality assurance of high product qualities based on modelling and numerical simulations. Possibilities include: temporal heating of the deposit within the spray phase (minimization of porosity effects within the bottom layers of the deposit), or successive cooling or heating of specific areas of the deposit during the spray and the cooling phases especially for high-volume products such as billets. The potential of these process control mechanisms may be further investigated by simulation in spray forming applications.

6.4.6   Grain coarsening

Besides analysis of the microstructure of the spray particles upon impact, coarsening of the fine-grained solid–liquid microstructure in the mushy zone needs to be analysed. It has been observed (Annavarapu and Doherty, 1995) that segregate spacings after coarsening were smaller than predicted by empirical correlations of dendrite arm spacing and freezing time.

The substructure size in conventionally cast alloy is dependent on the amount of coarsening the dendrite arms undergo during solidification. Dendrite arm coarsening is a surface-tension-driven process. The solid at highly curved surfaces is melted and preferentially redeposited at sites with less curvature by solute transport through the liquid. Several models have been proposed for this process in which the smaller dendrite arms melt back leading to an overall increase in the average dendrite arm spacing. Simple analytical treatment, assuming diffusion-control of the coarsening process at low to medium solid fractions,

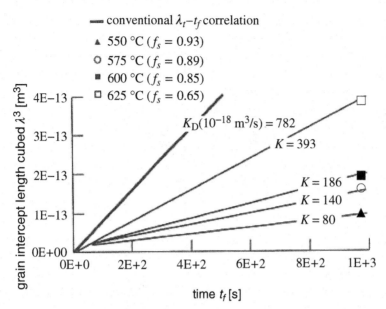

Fig. 6.61 Segregate spacing as a function of isothermal coarsening time for spray cast AA2014 Al alloy (Annavarapu and Doherty, 1995)

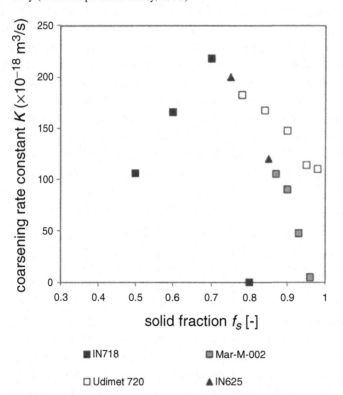

Fig. 6.62 Coarsening rate constant versus solid fraction for spray formed Ni and Cu alloys (Manson-Whitton *et al.*, 2002)

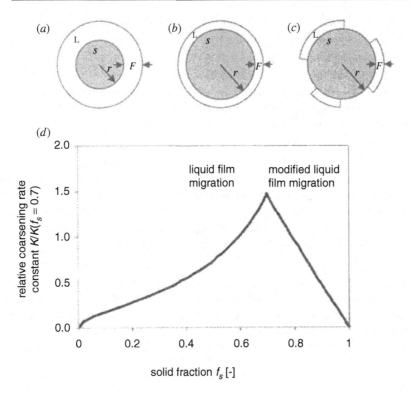

Fig. 6.63 Schematic of a solid grain of radius r surrounded by a liquid film of thickness F: (a) for $f_s < f_{s0}$, (b) for $f_s = f_{s0}$ and (c) for $f_s > f_{s0}$; (d) predicted coarsening rate constant as a function of solid fraction for the modified liquid film migration model for $f_{s0} = 0.7$ (Manson-Whitton *et al.*, 2002)

yields a typical cubic dependence of the dendrite arm spacing λ_t on the local solidification time t_f:

$$\lambda_t^3 = \lambda_0^3 + K_D t_f, \tag{6.37}$$

where K_D is the dendritic coarsening rate constant, which is a function of several material properties. The resulting $\lambda_t - t_f$ correlation is widely used to determine the cooling rate from the measured final spacings (dendrite arm, cell or segregate).

An important difference between spray formed materials and, for example, convention-ally cast materials is that in spray forming the grain coarsening process in the mushy zone is initiated at an already high solid content f_s. This leads to remarkable differences in the coars-ening rate constant K. The measured dependence of segregate spacing λ_t (Annavarapu and Doherty, 1995) on isothermal coarsening time t_f for spray deposited AA2014 aluminium alloy is illustrated in Figure 6.61. The graph shows a reduction in the coarsening rate at lower temperatures, i.e. at higher solid fractions f_s. Also shown is the conventional corre-lation of segregate spacing on solidification time. The samples had an originally equiaxed microstructure with an average grain size of 17 μm. The observed segregate spacings after coarsening were smaller than predicted by the empirical correlation. The coarsening was found to become slower as the temperature was reduced and the solid fraction f_s increased. The conventional coarsening theory and experiments predict the opposite, i.e. faster coarsening at higher volume fractions of solid. Two models have been developed for grain growth at high volume fractions of solid by processes whose rates are limited by migration of liquid at grain boundaries as liquid films (two-grain contact surface) or liquid rods (three-grain contacts at triple points). The conventional diffusion limited coarsening law was reproduced, but the rate constant K contained the term $(1/(1 - f_s))$ and so also predicted accelerated coarsening at $f_s \to 1$.

Manson-Whitton *et al.* (2002) have shown that for spray formed materials at higher solid fractions, the coarsening rate increases with f_s for $f_s < 0.75$ and then decreases again with further increasing f_s for $f_s > 0.75$, as illustrated in Figure 6.62, where the cubic coarsening rate constant as a function of solid fraction for spray formed Ni and Cu alloys is to be seen. A model for liquid film migration is proposed which takes into account the reducing area of the liquid film as f_s increases for high solid fraction values. This effect is illustrated in Figure 6.63, where a schematic of a solid grain of radius r, surrounded by a liquid film of constant thickness F for (*a*) $f_s < f_{s0}$, (*b*) $f_s = f_{s0}$ and (*c*) $f_s > f_{s0}$: reduction of the solid area in contact with the liquid is shown. In Figure 6.63(*d*) the predicted coarsening rate constant as a function of solid fraction for the modified liquid film migration model is illustrated for $f_{s0} = 0.7$. The formation of intergranular liquid droplets and the pinning of grain boundary liquid films by dispersoids during coarsening has been discussed.

7 An integral modelling approach

Coupling of several mechanisms into an integrated model for the spray forming process is the final aim of simulation. Such an integral spray forming model has been investigated, for example, by Bergmann (2000), Minisandram *et al.* (2000) (which has already been introduced in Section 6.3) and Pedersen (2003).

Connection between those submodels aforementioned has been performed, for transient temperature material behaviour, from melt superheating in the tundish to room temperature in the preform via cooling and solidification, in a three-stage approach in Bergmann (2000). Tundish melt flow and a thermal model are accounted for in the first stage. The local separation method is employed in the second stage to determine temperature-averaged properties of the spray and solid fractions in the particle mass at the centre-line of the spray. The temporal behaviour of a melt element is derived in terms of the averaged mean residence time of the particle mass. By combination of these data with calculated temporal cooling and solidification distributions of a fixed volume element inside the deposit, one yields the mean thermal history of the material at a specified location in the deposit.

The transient thermal and solidification distributions are shown in Figures 7.1 and 7.2. Results are illustrated separately for the three different modelling areas (with different time scales) for: (a) melt flow in the tundish, (b) particle cooling in the spray, (c) growth and cooling of the deposit. The process conditions used are for a steel spray forming of a Gaussian-shaped deposit (see Table 6.5).

From the top of the tundish (upper melt surface) to its exit in the atomization nozzle, the melt is cooled by approximately 20 K (for an initial melt temperature $T_L = 1933$ K). The solid fraction of the melt does not change as the melt stays fully liquid ($f_s = 0$). In total, the mean residence time of the melt in the tundish is 51.6 s for a tundish exit diameter $d_{exit} = 4$ mm and a surface height $h = 0.25$ m (determining the melt mass flow rate). Within the tundish a velocity profile develops in the flowing melt, with highest velocity values at the centre-line.

In the spray, convective heat transfer to the gas is mainly responsible for cooling of melt droplets. Until the spray impinges onto the substrate ($z = 0.6$ m; $s = 0.47$ m), the mean residence or flight time of the particles in the spray is 11 ms. During this time, the melt material cools down by $\Delta T = 260$ K ($T_m = 1650$ K) corresponding to a mean cooling rate of 2.4×10^4 K/s. Hereby the mean solid fraction in the spray increases to $f_s = 0.71$.

Because of thermal diffusion and the release of latent heat inside the deposit, the temperature of the material directly after compaction at first increases due to the further increase in quantity of the local solid fraction. For the conditions discussed, the temperature increases

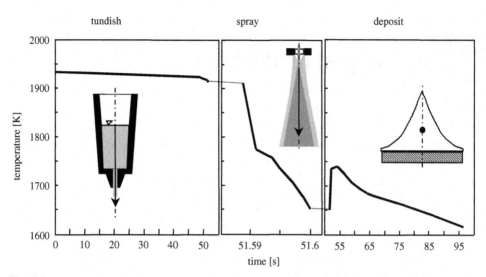

Fig. 7.1 Calculated mean material cooling in the three process steps. Calculation parameters are given in Table 6.5 (Bergmann, 2000)

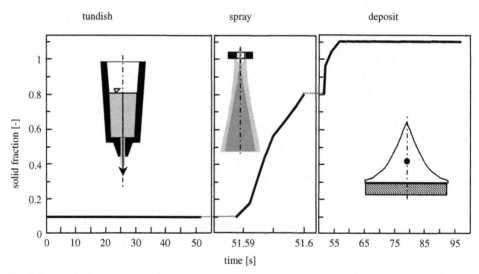

Fig. 7.2 Calculated mean solid fraction in the three process steps. Calculation parameters are given in Table 6.5 (Bergmann, 2000)

by approximately 90 K. After complete solidification of the volume element under investigation ($f_s = 1$), the temperature decreases again at a cooling velocity of approximately 2 to 3 K/s.

By variation of the standard process conditions, the thermal properties of the particle mass in the spray are responsible for the change in transient cooling and solidification processes inside the deposit, as shown in Figure 7.3. Here increase of the spray distance

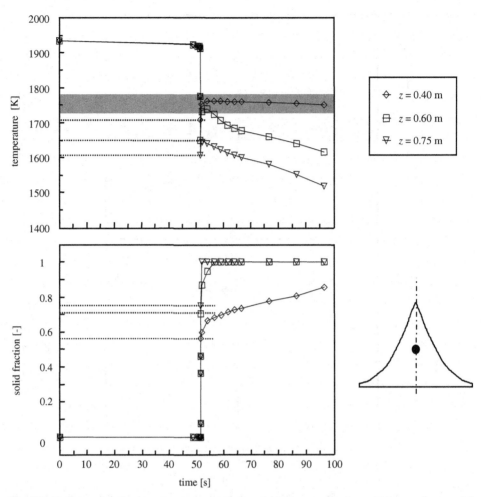

Fig. 7.3 Influence of distance z between atomizer and substrate on the thermal history of a material element. Calculation parameters are given in Table 6.5 (Bergmann, 2000)

between atomizer and substrate is illustrated. For a distance of $z = 0.4$ m, the compacting particle mass has a mean temperature of $T_m = 1707$ K and a mean solid fraction of $f_s = 0.56$. Due to the high content of liquid melt remaining, the volume element (discussed here at the end of the time shown), is not completely solidified ($f_s = 0.86$). Thus the temperature only decreases slightly, by approximately 12 K, within the solidification interval.

By increasing the spray distance to 0.6 m the mean thermal conditions of the impacting mass are changed to $T_m = 1650$ K and $f_{s,m} = 0.71$. The temperature of the material reaches the solidification interval for a short time and further decreases once the solidification process has finished. At a spray distance of $z = 0.75$ m, the particle mass completely solidifies immediately after compaction. After the temperature rises towards the solidification interval, the material cools down continuously. In comparison to the other two spray distances, the material inside the deposit cools down fastest in this case.

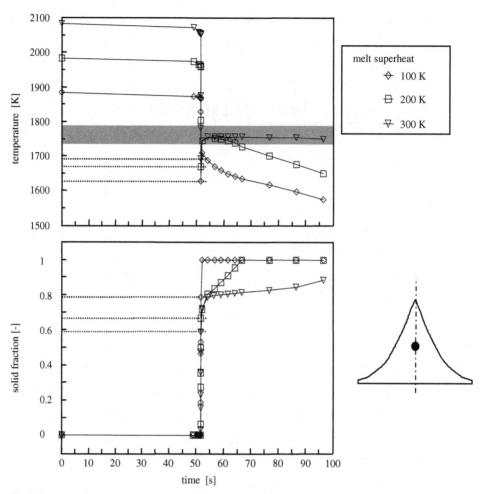

Fig. 7.4 Influence of melt superheating on the thermal history of the material. Calculation parameters are given in Table 6.5 (Bergmann, 2000)

In another parametric study, the amount of melt superheating in the tundish is changed. Augmentation of the melt superheat (from 100 to 300 K, parameter settings P1 to P3) results in higher mean temperatures and an increased amount of melt remaining in the impacting particle mass, as seen in Figure 7.4. For a melt superheat of $\Delta T_L = 300$ K, the amount of liquid melt remaining in the particles is sufficient for the material inside the deposit to be only partially solidified at the time scale shown here. The temperature is retained after reheating in the solidification interval. For a melt superheat of $\Delta T_L = 200$ K, the volume element discussed here is completely solidified after 66 s. After increasing during compaction, the temperature of the melt reaches a maximum of $T = 1752$ K at $t = 54$ s and afterwards decreases. A melt superheat of just $\Delta T_L = 200$ K results in sudden solidification of the material after compaction. The latent heat released is too low to increase the temperature

Table 7.1 *Spray forming parameters for coupled simulation. Values in italics are for adaptation of boundary conditions (Pedersen, 2003)*

Melt mass flow rate $\dot{M}_l$ [kg/s]	0.2
Gas mass flow rate $\dot{M}_g$ [kg/s]	0.2–0.6
Deposit rotational velocity ω [1/min]	116
Spray angle α [°]	35
Eccentricity e [m]	0.02
Distance to atomizer z [m]	0.5
Lubanska constant [-]	50
Melt stream diameter d_l [mm]	*3.7/7.4*
Total area of gas nozzles A_0 [m^2]	*0.0003/0.0012*

of the material above the liquidus. After a short increase of the temperature, the material cools down immediately.

An integrated numerical model of the spray forming process has been developed by Pedersen (2003). The model is able to predict the shape and temperature of the spray formed preform and takes into account such factors as thermal coupling between the gas and the droplets, the change in droplet-size distribution radially in the spray, the shadowing effect exhibited by billet-shaped preforms and the sticking efficiency of Gaussian-shaped preforms. The model has been used to derive a relationship between the gas to melt mass flow ratio GMR and surface temperature for the billet shape during spraying. The surface temperature is an important process control parameter in spray forming applications. It depends directly on the energy, respectively, the size, temperature and solid fraction, contained in the particles upon impact.

For variation of the GMR, the melt mass flow is chosen and the proper gas flow rate is adjusted. The melt mass flow rate is varied from 0.05 to 0.2 kg/s. The case of a melt mass flow rate of 0.2 kg/s is discussed here. The process parameters investigated within spray forming of billets of 100Cr16 in this case are listed in Table 7.1. Figure 7.5 illustrates the calculated billet shape at different time steps. The billet produced is rather small, with a width of 0.1 m and a height of 0.3 m after a spraying time of 60 s. An almost stationary billet shape is achieved after approximately 20 s spray time in this case; afterwards the shape of the top of the billet does not change, but the billet grows in height. The spray model shows that for varying melt mass flow rates, simply adjusting the GMR will not achieve plausible results. Figure 7.6 shows the calculated solid fraction in the spray as a function of distance from the atomizer and the melt mass flow rate for a constant GMR of 2. The prescribed ideal spray impingement distance of 0.5 to 0.6 m is indicated in the figure. In most investigations of the spray forming process, it has been found that the desirable solid fraction in the spray upon impact is between 50 and 70%. Therefore, a lower limit of 45% and a higher limit of 75% are indicated in the plot to define the region within which the solid fraction should be achieved too. As can be seen, the calculated solid fraction within the impact region gives reasonable results only for the lowest melt mass flow rate under investigation, i.e. for

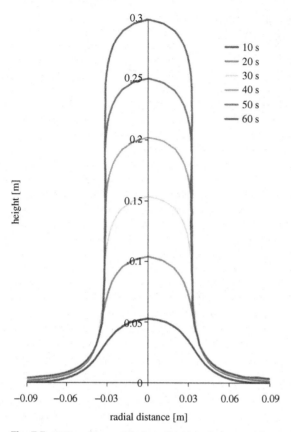

Fig. 7.5   Billet-shape evolution with time for a mass flow rate of 0.2 kg/s (Pedersen, 2003)

0.05 kg/s. Increased melt mass flow rates result in unrealistically high amounts of solid fractions upon impingement. In order to overcome this deficit a modification has been used at higher melt flow rates, which adjusts some atomizer design parameters to the given change in mass flow. This is the melt stream diameter and the area of gas delivery nozzles. By applying these changes to the model, the solid fractions in the spray upon impact fall in a realistic range, as can be seen for the as-modified cases indicated in the figure.

   Based on this reconfiguration of the atomizer, the proper GMR range for successful spray forming at each melt mass flow rate can be calculated. The result for a 0.2 kg/s melt flow rate is illustrated in Figure 7.7, where a wide range of GMR values from 1 to 3 has been found to achieve spray solid fractions between 45 and 75%. The overall mass median drop-size distribution in the spray dependent on GMR has been taken from Lubanska's (1970) formula, Eq. (4.49), and the radial dependency of the mean drop size in the spray is calculated from experimental observations (Hattel *et al.*, 1999) as:

$$d_{50}^{\text{local}}(r) = d_{50}^{\text{const}}\, e^{(-r/0.621)}. \tag{7.1}$$

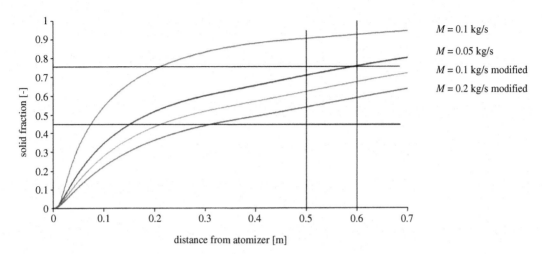

Fig. 7.6   Average solid fraction of the spray as a function of distance from the atomizer and melt mass flow rate GMR = 2. Data for standard and modified parameters are presented in Table 7.1 (Pedersen, 2003)

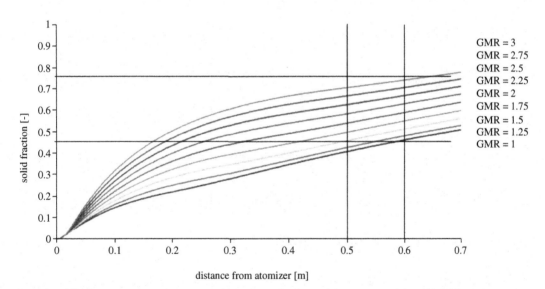

Fig. 7.7   Average solid fraction of the spray as a function of distance from atomizer and GMR, for a constant melt mass flow rate of 0.2 kg/s. Calculation parameters are given in Table 7.1 (Pedersen, 2003)

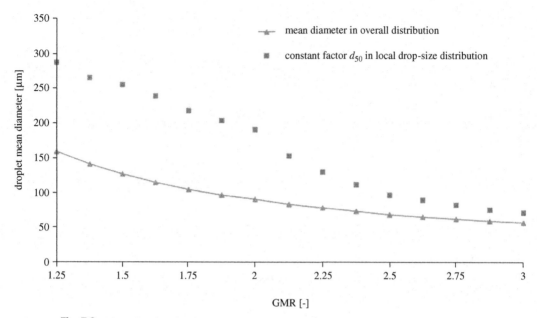

Fig. 7.8 Mean droplet diameter and local drop-size distribution coefficient as function of GMR for a constant melt mass flow rate of 0.2 kg/s (Pedersen, 2003)

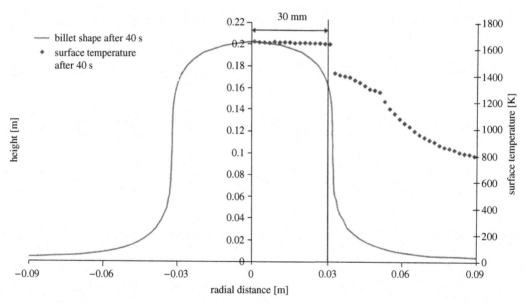

Fig. 7.9 Billet shape and surface temperature after 40 s spray time for GMR = 1.5 and a melt mass flow rate of 0.2 kg/s (Pedersen, 2003)

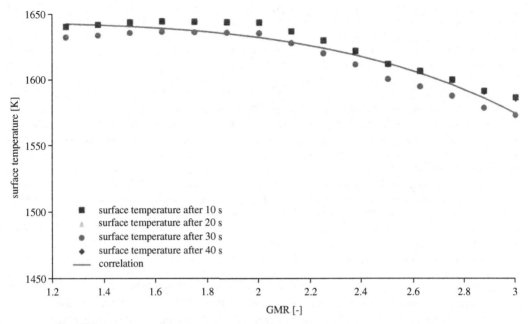

Fig. 7.10   Surface temperature as a function of GMR using $a = 1644.57$, $b = 4.51$, $x_0 = 6.02$ and a constant melt mass flow rate of 0.2 kg/s (Pedersen, 2003)

This correlation contains a diameter coefficient depending on the GMR and fulfilling the integral condition given by the overall mass median spray diameter. The result for the local mean drop size in the spray is shown in Figure 7.8. The standard deviation of the drop-size distribution has been taken from:

$$\sigma = \sqrt[3]{d_{50.3}/13}. \tag{7.2}$$

The main result of the next modelling step, the billet model, is the billet surface temperature distribution. The calculated surface temperature on the billet after a spray time of 40 s is illustrated in Figure 7.9 for the case of GMR = 1.5 and a melt mass flow rate of 0.2 kg/s. The time of 40 s has been chosen because at that time a quasi-stationary state is reached. Two regions on the billet surface may be distinguished. In the centre of the billet, up to $r = 30$ mm, the surface temperature is almost constant. In the outer region of the billet, $r > 30$ mm, the surface temperature decreases to about half the value at the centre. Because the surface temperature at the centre of the billet is almost constant, an empirical correlation in this area has been derived from this calculation.

The calculated surface temperature in the centre of the billet is dependent on GMR and is illustrated in Figure 7.10. The surface temperature decreases with increasing GMR, while at lower GMR values almost constant surface temperatures are achieved. The decrease in surface temperature is due to the decreasing solid fraction in the spray with increasing GMR because of the decreasing mean drop size in the spray. Correlation between the GMR and the surface temperature is assumed to be given in the form:

$$T_{surface} = \frac{a}{\left[1 + \left(\frac{GMR}{x_0}\right)^b\right]}. \tag{7.3}$$

The relationships for the factors in this equation are derived for a GMR for which the atomizer provides a solid fraction within the desirable range of 45 to 75%. The upper and lower bounds of this range are GMR_{high} and GMR_{low} and the respective surface temperatures at these bounds need to be known. For the constants a and b the correlations are:

$$a = T(GMR_{low}), \tag{7.4}$$

$$b = \frac{T(GMR_{low}) - T(GMR_{high})}{10\,(GMR_{high} - GMR_{low})} + (GMR_{high} - GMR_{low}), \tag{7.5}$$

and x_0 represents a reference GMR:

$$x_0 = GMR_{high}^2 - (GMR_{high} + GMR_{low}). \tag{7.6}$$

In the last step of Pedersen's (2003) integrated modelling approach, the temperature distribution inside the sprayed billet is discussed. The results of the temperature distribution within the spray formed product are in agreement with the discussions in Section 6.3.

8   Summary and outlook

The influence and importance of numerical models and simulations in science and engineering as appropriate tools for:

- analysis of engineering processes, as well as for
- conception and design of processes, and the
- development and analysis of control mechanisms,

has rapidly increased in recent years. In some technical research and development areas, simulation has been employed as an important contribution, given identical ranking as experiment and theory. This influence is valid not only in universities and research laboratories, but also in industry.

This technical progress of numerical simulation tools is based on ongoing rapid developments that have been achieved in hardware and numerics, and also on some important developments in modelling of physical and technical processes. These models can be incorporated and implemented into simulation codes that become easy to use. In recent developments in this area, similar success compared to experimental or physical measurement techniques have been achieved.

The possibility of using a simulation model to decouple some of the physical effects and mechanisms involved in a complex technical process, which may only be sequentially analysed by experimental means, highlights the potential of this new analytical approach. Here physical understanding of complex processes may be derived and used to optimize and develop processes. This contributes not only to scientific understanding, but also to economic and ecological technical innovations.

As an example of modelling and numerical process simulation, in this book fluid atomization processes and the spray forming of metals have been investigated, with particular reference to transport and exchange processes within multiphase flow, including momentum, heat and mass transfer.

Spray forming of metals for preform production is an impact orientated spray process. Spray forming is a metallurgical process for production of near-net shaped preforms with outstanding material properties. This process has recently been developed from laboratory to industrial scale. Here, as in some other technical developments, scientific and physical understanding lags behind realization of the process. The potentials of simulation may be developed in parallel to practical realizations. Modelling and simulation has added an important contribution to further spray forming process developments.

Each technical process may be further divided into a sequential number of subprocesses. Integral modelling of the complete process requires adapted division of the process. Within spray forming these main subprocesses can be defined as:

- fragmentation and atomization of the metal melt,
- dispersed multiphase spray flow,
- compaction of the spray and realization and growth of the deposit.

These subprocesses may once again be subdivided into individual tasks. These tasks then have to be individually modelled, based on physical descriptions, and finally analysed. Thereafter, these submodels need to be coupled to result in an integral process model. Modelling and simulation of several spray forming tasks are presented here. The individual basis for each is derived and from here the simulation tools are outlined. Application of these models with respect to the spray forming process is discussed and related to experiments and measurements, where data are available.

At the very least, this book gives a thorough introduction to a new analysis tool for processes within spray applications. This approach has been tailored here to the spray forming process discussed. Necessary steps for further integration of submodels into more complex integral spray process models have been outlined and some have already been realized theoretically. Interaction between submodels and their integration into an integral spray forming process model may be generalized in the future.

Bibliography

Aamir, M. A. and Watkins, A. P. Dense propane spray analysis with a modified collision model, CD-ROM, *Proc. ILASS-Europe'99*, 5–7 July Toulouse (1999)

Abbott, C. E. A survey of water drop interaction experiments, *Rev. Geophys. Space Phys.* **15** (1977): 363–74

Ahmadi, M. and Sellens, R. W. A simplified maximum-entropy-based drop size distribution, *Atomiz. & Sprays* **3** (1993): 292–310

Ahrens, O. Numerische Simulation des transsonischen Strömungsfeldes von unterexpandierten Freistrahlen, Studienarbeit, Fachgebiet Verfahrenstechnik, Universität Bremen (1995)

Albrecht, A., Bedat, B., Poinsot, T. J. and Simonin, O. Direct numerical simulation and modeling of evaporating droplets in homogeneous turbulence: application to turbulent flames, CD-ROM, *Proc. ILASS-Europe'99*, 5–7 July, Toulouse (1999)

Amsden, A. A., O'Rourke, P. J. and Butler, T. D. *KIVA-II: A Computer Program for Chemically Reactive Flows with Sprays*, Report LA-11560-MS, Los Alamos National Laboratory, LNM (1989)

Andersen, O. Berechnung des Temperaturverlaufs einer stationären Schmelzeströmung durch ein dünnes Rohr mit Kreisquerschnitt unter gleichzeitiger Berücksichtigung von Strahlung und Konvektion, Diplomarbeit, Fachgebiet Verfahrenstechnik, Universität Bremen (1991)

Anderson, I. E. and Figliola, R. S. Observations of gas atomization process dynamics, in: P. U. Gummeson and D. A. Gustafson (eds.) *Modern Developments in Powder Metallurgy*, Vol. 20, Metal Powder Industries Federation, Princeton, NJ (1988), pp 205–23

Anderson, I. E. and Terpstra, R. L. Progress toward gas atomization processing with increased uniformity and control, *Mater. Sci. Engng.* **A326** (2002) 1: 101–9

Anderson, I. E., Terpstra, R. L. and Rau, S. Progress toward understanding of gas atomization processing physics, *Kolloquium des SFB 372*, vol. 5, Universität, Bremen (2001) pp 1–16

Annavarapu, S. and Apelian, D. and Lawley, A. Processing effects in spray casting of steel strip, *Metall. Trans. A* **19** (1988): 3077–86

Annavarapu, S., Apelian, D. and Lawley, A. Spray casting of steel strip – process analysis, *Metall. Trans. A* **21** (1990): 3237–56

Annavarapu, S. and Doherty, R. D. Evolution of microstructure in spray casting, *Int. J. Powder Metall.* **29** (1993) 4: 331–43

Annavarapu, S. and Doherty, R. D. Inhibited coarsening of solid-liquid microstructures in spray casting at high volume fractions of solid, *Acta Metall. Mater.* **43** (1995) 8: 3207–30

Anno, J. N. *The Mechanics of Liquid Jets*, Lexington Books, Lexington, MA (1977)

Antipas, G., Lekakou, C. and Tsakiropoulos, P. The break-up of melt streams by high pressure gases in spray forming, *Proc. 2nd International Conference on Spray Forming ICSF-2*, 13–15 Sept., Swansea (1993) pp 15–24

Armster, S. Q., Delplanque, J. P., Rein, M. and Lavernia, E. J. Thermo-fluid mechanisms controlling droplet based materials processes, *Int. Mater. Rev.* **47** (2002) 6: 265–301

Ashgriz, N. and Poo, J. Y. Coalescence and separation in binary collision of liquid drops, *J. Fluid Mech.* **221** (1990): 183–204

Baehr, H. D. and Stephan, K. *Wärme- und Stoffübertragung*, Springer-Verlag, Berlin (1994)

Barrett, R., Berry, M., Chan, T. F., Demmel, J., Donato, J., Dongarra, J., Eijkhout, V., Pozo, R., Romine, C. and van der Vorst, H. *Templates for the Solution of Linear Systems: Building Blocks for Iterative Methods*, 2nd Edition, SIAM Publishing, Philadelphia, PA (1994). Available at: ftp.netlib.org/templates/templates.ps

Bauckhage, K. Das Zerstäuben metallischer Schmelzen, *Chem.-Ing.-Tech.* **64** (1992) 4: 322–32

Bauckhage, K. Stand der Technik beim Sprühkompktieren von Bolzen, *Härt.-Tech.-Mitteilung* **52** (1997) 5: 319–31

Bauckhage, K. Use of the phase-Doppler-anemometry for the analysis and the control of the spray forming process, *Proc. PM²TEC'98*, 31 May–4 June, Las Vegas (1998a)

Bauckhage, K. Die Bedeutung der Partikelabkühlung für den Materialaufbau beim Sprühkompaktieren, *Kolloquium des SFB 372*, Vol. 4, Universität Bremen (1998b) pp 139–74

Bauckhage, K., Bergmann, D., Fritsching, U., Lohner, H., Schreckenberg, P. and Uhlenwinkel, V. Das Scaling-Down-Problem bei der Zweistoffzerstäubung von Metallschmelzen, *Chem.-Ing.-Tech.* **73** (2001) 4: 304–13

Bauckhage, K., Bergmann, D. and Tillwick, J. Die Massen- und Enthalpiebilanzierung des Sprühkegels als Kopplung für die Modellvorstellung des Materialaufbaus in der Mix-Schicht, *Kolloquium des SFB 372*, Vol. 4, Universität Bremen (1999) pp 139–70

Bauckhage, K. and Fritsching, U. Production of metal powders by gas atomization, in: Cooper, K. P., Anderson, I. E., Ridder, S. D. and Biancanello F. S. (eds.) Liquid Metal Atomization: Fundamentals and Practice, June TMS, Warrendale (2000) pp 23–36

Bauckhage, K., Liu, H. M. and Fritsching, U. Models for the transport phenomena in a new spray compacting process, 4th *Proc. International Conference on Liquid Atomization and Spray Systems, ICLASS'88*, Sendai/Japan, 21–24 August, The Fuel Society of Japan, Tokyo (1988) pp 424–30

Bauckhage, K. und Uhlenwinkel, V. Zu den Möglichkeiten eines automatisierten und optimierten Sprühkompaktierbetriebes, *Härt.-Tech.-Mitteilung* **51** (1996b) 5: 289–97

Bauckhage, K. und Uhlenwinkel, V. (eds.) Sprühkompaktieren – Sprayforming, *Kolloquium des SFB 372*, Vol. 1, Universität Bremen (1996a)

Bauckhage, K. und Uhlenwinkel, V. (eds.) Sprühkompaktieren – Sprayforming, *Kolloquium des SFB 372*, Vol. 2, Universität Bremen (1997)

Bauckhage, K. und Uhlenwinkel, V. (eds.) Sprühkompaktieren – Sprayforming, *Kolloquium des SFB 372*, Vol. 3, Universität Bremen (1998)

Bauckhage, K. und Uhlenwinkel, V. (eds.) Sprühkompaktieren – Sprayforming, *Kolloquium des SFB 372*, Vol. 4, Universität Bremen (1999)

Bauckhage, K. und Uhlenwinkel, V. (eds.) Sprühkompaktieren – Sprayforming, *Kolloquium des SFB 372*, Vol. 5, Universität Bremen (2001)

Bauckhage, K., Uhlenwinkel, U. and Fritsching, U. (eds.) *Proc. Spray Deposition and Melt Atomization Conference SDMA 2000*, 26–28 June, Bremen, Universität Bremen (2000)

Bauckhage, K., Fritsching, U., Uhlenwinkel, U., Ziesemis, J. and Leatham, A. (eds.) *Proc. 2nd Spray Deposition and Melt Atomization Conference SDMA 2003*, 22–25 June, Bremen, Universität Bremen (2003)

Baum, S. *Software for Graphics and Data Analysis*, Deptartment of Oceanography, Texas A&M University (1996). Available at: http://www-ocean.tamu.edu/~baum/ocean_graphics.html

Bayvel, L. and Orzechowski, Z. *Liquid Atomization*, Taylor & Francis, Washington, DC (1993)

Bellan, J. Perspectives on large eddy simulations for sprays: issues and solutions, *Atomiz. & Sprays* **10** (2000): 409–25

Beretta, F., Cavalieri, F. and D'Alessio, A. Drop size concentration in a spray by sideward laser light scattering measurements, *Combust. Sci. Technol.* **36** (1984): 19–37

Berg, J. C. (ed.) *Wettability*, Marcel Dekker, New York (1993)

Berg, M. Zum Aufprall, zur Ausbreitung und Zerteilung von Schmelzetropfen aus reinen Metallen, Dissertation Universität Bremen (1999)

Berg, M. and Ulrich, J. Experimental based detection of the splash limits for the normal and oblique impact of molten metal particles on different substrates, *J. Mater. Synth. Proc.* **5** (1997) 1: 45–9

v. Berg, E., Bürger, M., Cho, S. H. and Schatz, A. Analysis of atomization of a liquid jet taking into account effects of the near surface boundary layer, *Proc. 11th ILASS-Europe Conference* 21–23 March, Nürnberg (1995)

Bergmann, D. Modellierung des Sprühkompaktierprozesses für Kupfer- und Stahlwerkstoffe, Dissertation, Universität Bremen (2000)

Bergmann, D., Bauckhage, K. and Fritsching, U. Modelling the spray forming process, *1999 International Conference on Powder Metallurgy and Particulate Materials*, 20–24 June, Vancouver (1999b).

Bergmann, D., Fritsching, U. and Bauckhage, K. Modellierung der Abkühlung und raschen Erstarrung von Metalltropfen im Fluge während des Sprühkompaktierens, in: *Kolloquium des SFB 372*, Vol. 3, Bauckhage, K. and Uhlenwinkel, V. (eds.), Universität Bremen (1998) pp 175–96

Bergmann, D., Fritsching, U. and Bauckhage, K. Averaging thermal conditions in molten metal sprays, in: Mishra, B. (ed.) *Proc. TMS – Annual Meeting, EPD Congress 1999*, 28 Feb.–4 March, San Diego CA (1999a)

Bergmann, D., Fritsching, U. and Bauckhage, K. Coupled simulation of molten metal droplet sprays, in: Rath, H. J. (ed.) *8th International Symposium on Computational Fluid Dynamics ISCFD'99*, 5–10 September, Bremen (1999b)

Bergmann, D., Fritsching, U. and Bauckhage, K. A mathematical model for cooling and rapid solidification of molten metal droplets, *Int. J. Therm. Sci.* **39** (2000): 53–62

Bergmann, D., Fritsching, U. and Bauckhage, K. Simulation of molten metal droplet sprays, *Comp. Fluid Dynamics J.* **9** (2001a): 203–11

Bergmann, D., Fritsching, U. and Bauckhage, K. Thermische Simulation des Sprühkompaktierprozesses, *Härt.-Tech.-Mitteilung* **56** (2001b) 2: 110–19

Bergmann, D., Fritsching, U. and Crowe, C. T. Multiphase flows in the spray forming process, *Proc. 2nd International Conference on Multiphase Flow*, 3–7 April, Kyoto, Japan, Vol. 1 (1995) pp SP1–SP8

Bergström, C., Fuchs, L. and Holmborn, J. Large eddy simulation of spray injected in a strong turbulent cross flow, CD-ROM, *Proc. ILASS-Europe 99*, 5–7 July, Toulouse (1999)

Berthomieu, P., Carntz, H., Villedieu, P. and Lavergne, G. Characterization of droplet breakup regimes, in: Yule, A. J. (ed.) *Proc. ILASS-Europe'98*, 6–8 July, Manchester (1998) pp 72–7

Bewlay, B. P. and Cantor, B. Modeling of spray deposition – measurement of particle size, gas velocity, particle velocity, and spray temperature in gas-atomized sprays, *Metall. Trans. B* **12B** (1990): 899–912

Bhagat, R. B. and Amateau, M. F. Droplet solidification and microstructure modeling for Al–4Li alloy, *Adv. Powder Metall. & Parti. Mater.* **2** (1996)

Bird, R. B., Stewart, W. E. and Lightfoot, E. N. *Transport Phenomena*, Wiley International Edition, John Wiley & Sons, New York (1960)

Birtigh, A., Lauschke, G., Schierholz, W. F., Beck, D., Maul, C., Gilbert, N., Wagner, H.-G. and Werniger, C. Y. CFD in der chemischen Verfahrenstechnik aus industrieller Sicht, *Chem.-Ing.-Tech.* **72** (2000) 3: 175–93

Boettinger, W. J., Coriel, S. R., Greer, A. L., Karma, A., Kurz, W., Rappaz, M. and Trivedi, R. Solidification microstructures: recent developments, future directions, *Acta Mater.* **48** (2000) 1: 43–70

Brackbill, J. U., Kothe, D. B. and Zemach, C. A continuum method for modelling surface tension, *J. Comp. Phys.* **100** (1992): 335–54

Bradley, D. On the atomization of liquids by high-velocity gases, Part 1, *J. Phys. D: Appl. Phys.* **6** (1973a): 1724–36

Bradley, D. On the atomization of liquids by high-velocity gases, Part 2, *J. Phys. D: Appl. Phys.* **6** (1973b): 2267–72

Brander, B. and Brauer, H. *Impuls- und Stofftransport durch die Phasengrenzfläche von kugelförmigen fluiden Partikeln*, Fortschritt-Berichte VDI, vol. 3, No. 326, VDI-Verlag, Düsseldorf (1993)

Bricknell, R. H. The structure and properties of a nickel-base superalloy produced by Osprey atomization and deposition, *Metall. Trans A.* **17A** (1986): 583–91

Brody, H. D. and Flemings, C. Solute redistribution in dendritic solidification, *Trans. Metall. Soc. AIME* **236** (1966) 5: 615–23

Brooks, R. G., Moore, C., Leatham, A. G. and Coombs, J. S. The Osprey process, *Powder Metall.* **2** (1977): 100–2

Buchholz, M. Untersuchung des Sprühkompaktierverhaltens an sprühkompaktierten Bolzen, Dissertation, Universität Bremen (2002)

Buchholz, M., Uhlenwinkel, V., v. Freyberg, A. and Bauckhage, K. Specific enthalpy measurement in molten metal spray, *Mater. Sci. Engng.* **A326** (2002) 1: 165–75

Bürger, M., v. Berg, E., Cho, S. H. and Schatz, A. Fragmentation processes in gas and water atomization plants for process optimization purposes, Part 1: discussion of the main fragmentation processes, *Powder Metall. Int.* **21** (1989) 6: 10–15

Bürger, M., v. Berg, E., Cho, S. H. and Schatz, A. Analysis of fragmentation processes in gas and water atomization plants for process optimization purposes, Part 2: modelling of growth and stripping of capillary waves in parallel shear flow – the basic fragmentation mechanism, *Powder Metall. Int.* **24** (1992) 6: 32–8

Bürger, M., Schwalbe, W., Kim, D. S., Unger, H., Hohmann, H. and Schins, H. Two-phase description of hydrodynamic fragmentation processes within thermal detonation waves, *J. Heat Transfer* **106** (1984): 728–34

Bussmann, M., Aziz, S. D. and Chandra, S. Photographs and simulations of molten metal droplets landing on a solid surface, *J. Heat Transfer* **122** (2000): 422

Bussmann, M., Mostaghimi, J. and Chandra, S. On a three-dimensional volume tracking model of droplet impact, *Phys. Fluids* **11** (1999): 1406–17

Butzer, G. A. The production-scale spray forming of superalloys for aerospace applications, *J. Metals* **51** (1999) 4, web-edition: http://www.tms.org/pubs/journals/*J. Metals*/9904/butzer/butzer-9904.html

Cai, C. A modelling study for the design and control of spray forming, PhD thesis, Drexel University (1995)

Cai, W. D. and Lavernia, E. J. Modeling of porosity during spray forming, *Mater. Sci. Engng. A* **226–8** (1997): 8–12

Cai, C., Warner, L., Annavarapu, S. and Doherty, R. Modelling microstructural development in spray forming: experimental verification, in: Wood, J. V. *Proc. 3rd International Conference on Spray Forming*, Cardiff, 1996, Osprey Metals Ltd, Neath (1997)

Cappus, J. M. and German, R. M. (eds.) *Proc. 1992 Powder Metallurgy World Congress, Vol. 1: Powder Production and Spray Forming*, 21–26 June, San Francisco, CA (1992)

Carter, W. T., Benz, M.-G., Basu, A. K., Zabala, R. J., Knudsen, B. A., Forbes Jones, R. M., Lippard, H. E. and Kennedy, R. L. The CMSF process: the spray forming of clean metal, *J. Metals* **51** (1999) 4, web-edition: http://www.tms.org/pubs/journals/JOM/9904/Carter/Carter-9904.html

Chang, D.-H., Kang, S., Lee, E.-S. and Ahn, S. Analysis of transient heat conduction with phase change in a spray deposited body, in: Marsh, S. P. *et al.* (eds.), *Solidification 1998*, The Minerals, Metals & Materials Society, Warrendale, PA (1998) pp 497–508

Chao, B. T. Motion of spherical gas bubbles in a viscous liquid at large Reynolds numbers, *Phys. Fluids* **5** (1962) 1: 69–79

Chen, M. M., Crowe, C. T., Fritsching, U., Pien, S. J., *et al.* (eds.) *Transport Phenomena in Materials Processing and Manufacturing, Heat Transfer Division* – vol. 336, *Fluids Engng Division* – Vol. 240, The American Society of Mechanical Engineers ASME, New York (1996)

Cheng, C., Annavarapu, S. and Doherty, R. Modelling based microstructural control in spray casting, *Proc. 2nd International Conference on Spray Forming, ICSF-2*, 13–15 Sept., Swansea (1993)

Clift, R., Grace, J. R. and Weber, M. E. *Bubbles, Drops and Particles*, Academic Press, San Diego, CA (1978)

Coimbra, C. F. M. and Rangel, R. H. General solution of the particle momentum equation in unsteady Stokes flows, *J. Fluid Mech.* **370** (1998): 53–72

Colella, P. and Glaz, H. M. Efficient solution algorithmus for the Riemann problem for real gases, *J. Comp. Phys.* **59** (1985): 264–89

Computational Fluid Dynamics Services *CFX 4.1 Flow Solver User Guide*, Computational Fluid Dynamics Services, Harwell Laboratories Oxfordshire (1995)

Conelly, S., Coombs, J. S. and Medwell, J. O. Flow characteristics of metal particles in atomized sprays, *Metal Powder Rep.* **41** (1986): 9

Cousin, J. and Dumouchel, C. Effect of viscosity on the linear instability of a liquid sheet, *Atomiz. & Sprays* **6** (1996): 563–76

Cousin, J. and Dumouchel, C. Theoretical determination of spray drop size distribution, *Proc. International Conference on Liquid Atomization and Spray Systems ICLASS'97*, August, Seoul (1997) Part 1: Description of the Procedure, pp 788–95, Part 2: Applications, pp 796–803

Cousin, J., Yoon, S. J. and Dumouchel, C. Coupling of the classical linear theory and the maximum entropy formalism for the prediction of drop size distributions in sprays, application to pressure swirl atomizers, *Atomiz. & Sprays* **6** (1996) 5: 601 ff.

creare.x Inc. (Hrsg.) *FLUENT User's Manual*, Version 3.02, Hanover (1990)

Crowe, C. T. Modelling spray–air contact in spray drying systems, *Adv. in Drying* **1** (1980) 3: 63–99

Crowe, C. T. Challenges in numerical simulation of metal sprays in spray forming processes, *Kolloquium des SFB 372*, Vol. 2, Universität. Bremen (1997) pp 1–16

Crowe, C. T. Importance of multiphase coupling in modeling metal-droplet sprays, *Proc. Spray Deposition and Melt Atomization SDMA 2000*, Bremen (2000) pp 757–70

Crowe, C. T., Sharma, M. P. and Stock, D. E. The particle-source-in-cell method for gas droplet flow, *J. Fluids Engng.* **99** (1977): 325–32

Crowe, C. T., Sommerfeld, M. and Tsuji, Y. *Multiphase Flows with Drops and Particles*, CRC Press, Boca Raton, CA (1998)

Crowe, C. T., Troutt, T. R. and Chung, J. N. Numerical models for two-phase turbulent flows, *Ann. Rev. Fluid Mech.* **28** (1996): 11–43

Cui, C., Cao, F., Li, Z. and Li, Q. Modeling of the spray forming and solidification process of billets, *Proc. 4th International Conference on Spray Forming*, Baltimore, MD (1999)

Cui, C., Cao, F., Li, Z., Zhang, J. and Li, Q. Modeling of spray forming and solidification process of tubular products, *Proc. Spray Deposition and Melt Atomization SDMA 2000*, Bremen, (2000) pp 825–38

Cui, C., Fritsching, U., Schulz, A., Bauckhage, K. and Mayr, P. Control of cooling during spray forming of bearing steel billet, *Proc. Spray Deposition and Melt Atomization SDMA 2003*, Bremen, 22–25 June, (2003) pp 8.117–8.128

Cui, C., Li, Z. and Li, Q. Numerical simulation of heat and momentum transfer in spray forming process, *Proc. 1998 PM World Congress*, Vol. 1, 18–22 October, Grenada (1998) pp 555–60

Dash, S. M. and Wolf, D. E. Interactive phenomema in supersonic jet mixing problems, Part I: phenomenology and numerical modeling techniques, *AIAA Journal* **22** (1984) 7: 905–13

Delplanque, J. P., Lavernia, E. J. and Rangel, R. H. Analysis of in-flight oxidation during reactive spray atomization and deposition processing of aluminum, *J. Heat Transfer* **122** (2000): 126–33

Delplanque, J.-P., Lavernia, E. J. and Rangel, R. H. Multidirectional solidification model for the decription of micropore formation in spray deposition processes, *Numerical Heat Transfer, Part A* **30** (1996): 1–18

Delplanque, J. P. and Rangel, R. H. Simulation of liquid-jet overflow in droplet deposition processes, *Acta Mater.* **47** (1999) 7: 2207–13

Delplanque, J. P. and Sirignano, W. A. Boundary-layer stripping effects on droplet transcritical convective vaporization, *Atomiz. & Sprays* **4** (1994) 3: 325–49

Dielewicz, L. G., v. Berg, E. and Lampe, M. Computation of transsonic two-phase flow in liquid metal jet atomizers, CD-ROM *Proc. ILASS-Europe'99*, 5–7 July Toulouse (1999)

Djuric, Z. and Grant, P. S. Two dimensional simulation of liquid metal spray deposition onto a complex surface II: splashing and redeposition, *Modelling Simul. Mater. Sci. Eng.* **9** (2001): 111–27

Djuric, Z., Newberry, P. and Grant, P. S. Two dimensional simulation of liquid metal spray deposition onto a complex surface, *Modelling Simul. Mater. Sci. Eng.* **7** (1999): 553–71

Dobre, M. and Bolle, L. Theoretical prediction of ultrasonic spray characteristics using the maximum entropy formalism, in: Yule, A. J. (ed.) *Proc. ILASS-Europe'98*, 6–8 July, Manchester (1998) pp 7–12

Doherty, R. D., Annavarapu, S., Cai, C. and Warner Kohler, L. K. Modeling based studies for control and microstructure development in spray forming, *Kolloquium des SFB 372*, Vol. 2, Universität Bremen (1997) pp 45–78

Doherty, R. D., Cai, C. and Warner-Kohler, L. K. Modeling and microstructural development in spray forming, *Int. J. Powder Metall.* **33** (1997) 3: 50–60

Dombrowski, N. and Johns, W. R. The aerodynamic instability and disintegration of viscous liquid sheets, *Chem. Engng. Sci.* **18** (1963): 203–14

Domnick, J., Raimann, J., Wolf, G., Berlemont, A. and Cabot, M.-S. On-line process control in melt spraying using phase-Doppler anemometry, *Proc. International Conference on Liquid Atomization and Spray Systems ICLASS '97*, 18–22 August (1997) Seoul

Drezet, J.-M. Thermomechanical aspects in solidification processes, *Kolloquium des SFB 372*, Vol. 3, Universität Bremen (1998) pp 53–82

Duda, J. L. and Vrentas, J. S. Fluid mechanics of laminar liquid jets, *Chem. Engng. Sci.* **22** (1967): 855–73

Dumouchel, C. Problemes lies a la d ún pulverizateur mecanique – hydrodynamique de chambre et instabilite de nappe, Dissertation Université Rouen (1989)

Dunkley, J. J. Liquid metal atomization – a suitable case for investigation, in: Yule, A. J. (ed.) *Proc. ILASS-Europe'98*, 6–8 July, Manchester (1998) pp P1–P6

Durao, D. F. G. The application of laser anemometry to free jets and flames with and without recirculation, PhD thesis, University of London (1976)

Durst, F., Milojevic, D. and Schönung, B. Eulerian and Lagrangian predictions of particulate two-phase flows: a numerical study, *Appl. Math. Modelling* **8** (1984): 101–15

Dykhuizen, R. C. Review of impact and solidification of molten thermal spray droplets, *J. Thermal Spray Technol.* **3** (1994): 351–61

Dykhuizen, R. C. and Smith, M. F. Gas dynamic principles of spray, *J. Thermal Spray Technol.* **7** (1998) 2:, 205–12

Ebert, T., v. Buch, F. und Kainer, K. U. Sprühkompaktieren von Magnesiumlegierungen im Rahmen des SFB 390, *Kolloquium des SFB 372*, Vol. 3, Universität Bremen (1998) pp 9–30

Ebert, T., Moll, F. and Kainer, K. U. Spray forming of magnesium alloys and composites, *Proc. 3rd International Conference on Spray Forming*, Cardiff, 1996, Osprey Metals Ltd, Neath (1997) pp 177–85

Edwards, C. F. Formulating large-eddy simulations of dense multiphase flows, in: Yule, A. J. (ed.) *Proc. ILASS-Europe'98*, 6–8 July, Manchester (1998) pp P7–P16

Elghobashi, S. E., Abou-Arab, T. W., Rirk, M. and Mostafa, A. Prediction of the particle laden jet with a two-equation turbulence model, *Int. J. Multiphase Flow* **10** (1984): 697–710

Espina, P. I. Numerical simulation of atomization gas flow, *Kolloquium des SFB 372*, Vol. 4, Universität Bremen (1999) pp 127–38

Espina, P. I. and Piomelli, U. Numerical simulation of the gas flow in gas metal atomizers, *Proc. 1998 ASME – Fluids Engng Division*, Washington (1998a), FEDSM98–4901

Espina, P. I. and Piomelli, U. Study of the gas jet in a close-coupled gas metal atomizer, *AIAA* Aerospace Science Meeting, 12–15 June, Reno, NV, Paper 98–0959 (1998b)

Evans, R. W., Leatham, A. G. and Brooks, R. G. The Osprey preform process, *Powder Metall.* **28** (1985): 13–20

Faeth, G. M. Structure and atomization properties of dense turbulent sprays, *23rd Symposium on Combustion*, The Combustion Institute, Pittsburgh, PA (1990) pp 1345–52

Faragó, Z. Activities on liquid atomization at the Research Center Lampoldshausen of the German Aerospace Research Establishment, *Proc. International Conference on Liquid Atomization and Spray Systems ICLASS '97*, 18–22 August, Seoul (1997) pp 345–52

Faragó, Z. and Chigier, N. Morphological classification of disintegration of round liquid jets in a coaxial air stream, *Atomiz. & Sprays* **2** (1992): 137–53

Ferziger, J. H. and Peric, M. *Computational Methods for Fluid Dynamics*, Springer Verlag, Berlin (1996)

Fletcher, C. A. J. *Computational Techniques for Fluid Dynamics Part 1: Fundamental and General Techniques; Part 2: Specific Techniques for Different Flow Categories*, 2nd Edition, Springer Verlag (1991)

Flow Science, *Flow-3D User's Manual*, Flow Science, Santa Fe, CA (1998)

Ford, R. E. and Furmidge, C. G. L. Impact and spreading of spray drops on foliar surfaces, *Wetting* (*Soc. Chem. Ind.*) **25**: 417–32 (1967)

Forrest, J., Lile, S. and Coombs, J. S. Numerical modelling of the Osprey process, *Proc. International Conference on Spray Forming, ICSF-2*, 13–15 September, Swansea (1993)

Frigaard, I. A. Growth dynamics of spray-formed aluminium billets, Part 1: steady state crown shapes, *J. Mater. Proc. Manuf. Sci.* **3** (1994a): 173–93

Frigaard, I. A. Growth dynamics of spray-formed aluminium billets, Part 2: transient billet growth, *J. Mater. Proc. Manuf. Sci.* **3** (1994b): 257–75

Frigaard, I. A. Controlling the growth of alluminium spray-formed billets, *Kolloquium des SFB 372*, Vol. 2, Universität Bremen (1997) pp 29–44

Frigaard, I. A. Spray-forming of large diameter billets using twin atomizer system: basic features of spray-form growth dynamics, *Proc. Spray Deposition and Melt Atomization SDMA 2000*, Bremen (2000) pp 839–54

Fritsching, U. Modelling the spray cone behaviour in the metal spray forming process, momentum and thermal coupling in two-phase flow, *Phoenics J. Comp. Fluid Dynamics* **8** (1995) 1: 68–90

Fritsching, U. and Bauckhage, K. Die Bewegung von Tropfen im Sprühkegel einer Ein- und einer Zweistoffdüse, *Chem.-Ing.-Tech.* **59** (1987) 9: 744–5

Fritsching, U. and Bauckhage, K. Numerical investigations on the atomization of molten metals, *3rd International Phoenics-User Conference*, 28 August–1 September, Dubrovnik CHAM Ltd, London (1989)

Fritsching, U. and Bauckhage, K. Investigations on the atomization of molten metals: the coaxial jet and the gas flow in the nozzle near field, *PHOENICS J. Comp. Fluid Dynamics* **5** (1992) 1: 81–98

Fritsching, U. and Bauckhage, K. Lagrangian modelling of thermal and kinetic droplet/particle behaviour in the metal spray compaction process, *Proc. ILASS-93/CHISA-93*, 29 August–3 September, Prague (1993)

Fritsching, U. and Bauckhage, K. Zum Impuls- und Wärmetransport bei der Zerstäubung und anschließenden Kompaktierung von Schmelzen, *Chem.-Ing.-Tech.* **66** (1994a) 3: 380–2

Fritsching, U. and Bauckhage, K. Sprays and jets for metallurgical applications, *Proc. 7th Workshop on Two-Phase Flow Predictions*, 11–14 April, Erlangen (1994b)

Fritsching, U. and Bauckhage, K. Spray modelling in spray forming, in: Chen, M. M. and Crowe, C. T. (eds.) Multiphase Flow and Heat Transfer in Materials Processing, presented at International Mechanical Engineering Congress 94, 6–11 November, Chicago, *ASME-FED* **201** (1994c): 49–54

Fritsching, U. and Bauckhage, K. Thermal treatment and conditions of the deposit in spray forming applications, in: Chen, M. M. and Crowe, C. T. (eds.) Multiphase Flow and Heat Transfer in Materials Processing, presented at International Mechanical Engineering Congress 94, 6–11 November, Chicago, *ASME-Fluids Engng Division* **201** (1994d): 7–18

Fritsching, U. and Bauckhage, K. *Sprayforming of Metals*, Ullmann's Encyclopedia of Industrial Chemistry, 6th Edition, 1999 electronic release, Wiley VCH, Weinheim (1999)

Fritsching, U., Bergmann, D. and Bauckhage, K. Metal solidification during spray forming, *Proc. International Conference on Liquid Atomization and Spray Systems ICLASS'97*, 18–22 August, Seoul (1997a)

Fritsching, U., Bergmann, D., Heck, U. und Bauckhage, K. Modellierung und Simulation des Sprühkompaktierprozesses, in: Bauckhage, K. und Uhlenwinkel, V. (eds.) *Kolloquium des SFB 372*, Vol. 2, Universität Bremen (1997b)

Fritsching, U., Bergmann, D., Heck, U. and Bauckhage, K. Particle size distribution width in gas atomization of molten metals, *1999 International Conference on Powder Metallurgy and Particulate Materials*, 20–24 June, Vancouver (1999)

Fritsching, U., Liu, H. and Bauckhage, K. Numerical modelling in the spray compaction process, *Proc. 5th International Conference on Liquid Atomization and Spray Systems, ICLASS-91*, Gaithersburg, MD, NIST SP813 (1991) pp 491–8

Fritsching, U., Heck, U. and Bauckhage, K. The gas-flowfield in the atomization region of a free fall atomizer, *Proc. International Conference on Liquid Atomization and Spray Systems ICLASS'97*, 18–22 August, Seoul (1997)

Fritsching, U., Liu, H. and Bauckhage, K. Two-phase flow and heat transfer in the metal spray compaction process, *Proc. International Conference on Multiphase Flows '91*, 24–7 September, Tsukuba (1991)

Fritsching, U., Uhlenwinkel, V. and Bauckhage, K. Spreading of the spray cone for spray forming applications, *Proc. Powder Metallurgy World Congress, PM-93*, 12–15 July, Kyoto (1993)

Fritsching, U., Uhlenwinkel, V. and Bauckhage, K. (eds.) Selected papers from the International Conference on Spray Deposition and Melt Atomization, SDMA-2000, *Mater. Sci. Engng.* **A326** (2002) 1

Fritsching, U., Uhlenwinkel, V., Bauckhage, K. and Urlau, U. Gas- und Partikelströmungen im Düsennahbereich einer Zweistoffdüse, Modelluntersuchungen zur Zerstäubung von Metall-schmelzen, *Chem.-Ing.-Tech.* **62** (1990) 2: 146–7

Fritsching, U., Zhang, H. and Bauckhage, K. Thermal histories of atomized and compacted metals, *Proc. Powder Metallurgy World Congress PM-93*, 12–15 July, Kyoto (1993a)

Fritsching, U., Zhang, H. and Bauckhage, K. Modelling of thermal histories and solidification in the spray cone and deposit of atomized and compacted metals, *Proc. International Conference on Spray Forming, ICSF-2*, 13–15 September, Swansea (1993b)

Fritsching, U., Zhang, H. and Bauckhage, K. Numerical simulation of temperature distribution and solidification behaviour during spray forming, *Steel Research* **65** (1994a) 7: 273–8

Fritsching, U., Zhang, H. and Bauckhage, K. Numerical results of temperature distribution and solidification behaviour during spray forming, *Steel Research* **65** (1994b) 8: 322–5

Frohn, A. and Roth, N. *Dynamics of Droplets*, Springer Verlag, Berlin (2000)

Fukai, J., Shiiba, Y., Yamamoto, T., Miyatake, O., Poulikakos, D., Megaridis, C. M. and Zhao, Z. Wetting effects on the spreading of a liquid droplet colliding with a flat surface: experiment and modeling, *Phys. Fluids* **11** (1995): 236–47

Fukai, J., Zhao, Z., Poulikakos, D., Megaridis, C. M. and Miyatake, O. Modeling of the deformation of a liquid droplet impinging a flat surface, *Phys. Fluids* **5** (1993): 2588–99

Fukai, J., Asami, H. and Miyatake, O. Deformation and solidification behaviour of a molten metal droplet colliding with a substrate: modeling and experiment, in: Marsh, S. P. *et al.* (eds.) *Solidification 1998*, TMS, The Minerals, Metals & Materials Society, Warrendale, PA (1998), pp 473–83

Georjon, T. L. and Reitz, R. D. A drop-shattering collision model for multidimensional spray computations, *Atomiz. & Sprays* **9** (1999): 231–54

Gerking, L. Powder from metal and ceramic melts by laminar streams at supersonic speed, *Powder Metall. Int.* **25** (1993) 2: 59–65

Gerling, R., Liu, K. W. und Schimansky, F.-P. Pulverherstellung und Sprühformen von intermetallischen Titanbasislegierungen, *Kolloquium des SFB 372*, Vol. 4, Universität Bremen (1999) pp 105–26

Gerling, R., Schimansky, F. P., Wegmann, G. and Zhang, J. X. Spray forming of Ti 48.9Al (at%) and subsequent hot isostatic pressing and forging, *Mater. Sci. Engng.* **A326** (2002) 1: 73–8

Gosman, A. D. and Ioannides, E. Aspects of computer simulation of liquid-fueled combustors, *J. Energy* **7** (1983): 482–90

Grant, P. S., Cantor, B. and Katgerman, L. Modelling of droplet dynamic and thermal histories during spray forming. I. Individual droplet behaviour, *Acta Metall. Mater.* **41** (1993a) 11: 3097–108

Grant, P. S., Cantor, B. and Katgerman, L. Modelling of droplet dynamic and thermal histories during spray forming. II. Effect of process parameters, *Acta Metall. Mater.* **41** (1993b) 11: 3109–18

Grant, P. S. Spray forming, *Progress in Mater. Sci.* **39** (1995): 497–545

Grant, P. S. A model for the factors controlling spray formed grain sizes, *Kolloquium des SFB 372*, Vol. 3, Universität Bremen (1998) pp 83–92

Grant, P. S., Cantor, B. and Katgerman, L. *Acta Metall. Mater.* **41** (1993) 11: 3097

Grant, P. S., Underhill, R. P., Cantor, B. and Bryant, D. J. Modelling droplet behaviour during spray forming using FLUENT, TMS Annual Meeting, Orlando, FL (1997)

Grigull, U. and Sandner, H. *Wärmeleitung*, Springer-Verlag, Berlin, Heidelberg, New York, Tokyo (1986)

Gupta, M., Ibrahim, I. A., Mohammed, F. A. and Lavernia, E. J. Wetting and interfacial reactions in Al–Li–SiCp metal matrix composites processing by spray atomization and deposition, *J. Mater. Sci.* **26** (1991): 6673–84

Gupta, M., Mohammed, F. A. and Lavernia, E. J. Heat transfer mechanisms and their effects on microstructure during spray atomization and codeposition of metal matrix composites, *Mater. Sci. Engng.* **A144** (1991): 99–110

Gupta, M., Lane, C. and Lavernia, E. J. Microstructure and properties of spray atomized and deposited Al–7Si/SiC metal matrix composites, *Scripta Metall. Mater.* **26** (1992): 825–30

Gutierrez-Miravete, M., Lavernia, E. J., Trapaga, G. M. and Szekely, J. A mathematical model of the liquid dynamic compaction process. Part 2: formation of the deposit, *Int. J. Rapid Solidification* **4** (1988): 125–50

Hagerty, W. W. and Shea, J. F. A study of the stability of plane fluid sheets, *J. Appl. Mech.* **22** (1955): 509–14

Hansen, P. N., Hartmann, G. and Kallien, L. Numerical simulation of rapid solidification processes: powder and spray-forming technologies, *Solidification Processing* (1987): 373–6

Hansmann, S. und Müller, H. R. Hochzinnhaltige Bronzen mittels Sprühkompaktieren seigerungsarm hergestellt, *Kolloquium des SFB 372*, Vol. 4, Universität Bremen (1999) pp 1–6

Hardalupas, Y., Tsai, R.-E. and Whitelaw, J. H. Unsteady breakup of liquid jets in coaxial airblast atomizers, *Proc. International Conference on Liquid Atomization and Spray Systems ICLASS '97*, 18–22 August, Seoul (1997) pp 326–33

Hardalupas, Y., Taylor, A. M. K. P. and Wilkins, J. H. Experimental investigation of sub-millimetre droplet impingement onto spherical surfaces, *Int. J. Heat Fluid Flow* **20** (1999): 477–85

Harlow, F. H. and Shannon, J. P. The splash of a liquid drop, *J. Appl. Phys.* **38** (1967) 10: 3855–66

Hartmann, G. C. *Die Erstarrung von Metallen im Sprühgießprozeß am Beispiel der Zinnbronze CuSn6*, Fortschritt Berichte VDI, Reihe 5: Grund- und Werkstoffe No. 195, VDI-Verlag Düsseldorf (1990)

Hattel, J. H. Mathematical modelling and numerical simulation of casting processes, Technical University Denmark, Lyngby (1999)

Hattel, J. H., Pryds, N. H., Pedersen, T. B. and Pedersen, A. S. Numerical modelling of the spray forming process: the effect of process parameters on the deposited material, *Proc. Spray Deposition and Melt Atomization SDMA 2000*, Bremen 200 pp 803–812

Hattel, J. H., Pryds, N., Thorborg, J. and Ottosen, P. A quasi-stationary numerical model of atomized metal droplets, Part I: model formulation, *Modelling Simul. Mater. Sci. Engng.* **7** (1999) 3: 413–30

Heck, U. Zur Zerstäubung in Freifalldüsen, Dissertation, Universität Bremen (1998)

Heck, U., Fritsching, U. und Bauckhage, K. Zur Fluiddisintegration in Freifall-Zerstäubern, in: Koschel, W. W. and Haidn, O. J. (eds.) *Spray '97*, 3. Workshop über Sprays, Erfassung von Sprühvorgängen und Techniken der Fluidzerstäubung, DLR Lampoldshausen, 22–23 Oktober (1997)

Heck, U., Fritsching, U. and Bauckhage, K. Gas-flow effects on twin-fluid atomization of liquid metals, *Atomiz. & Sprays* **10** (2000) 1: 25–46

Helebrook, B. T. and Edwards, C. F. *Proc. 8th International Conference on Liquid Atomization and Spraying Systems ICLASS-2000*, 16–20 July, Pasadena, CA (2000)

Henein, H. Single fluid atomization through the application of impulses to a melt, *Mater. Sci. Engng.* **A326** (2002) 1: 92–100

Hetsroni, G. (ed.) *Handbook of Multiphase Systems*, Hemisphere, Washington, DC (1982)

Hill, J. M. *One-Dimensional Stefan Problems: An Introduction*, Longman Scientific & Technical, John Wiley, New York (1987)

Hinze, J. O. Fundamentals of the hydrodynamic mechanism of splitting in dispersion processes, *AIChE J.* **1** (1955): 289–95

Hirt, C. W., Nichols, B. D. and Romero, N. C. *SOLA – A Numerical Solution Algorithm for Transient Fluid Flows*, Report LA-5652, Los Alamos Scientific Laboratory, NM (1975)

Hirth, J. P. Nucleation, undercooling and homogeneous structures in rapidly solidified powders, *Metall. Trans. A*, **9A** (1978) 3: 401–4

Ho, S. and Lavernia, E. J. Thermal residual stresses in spray atomized and deposited Ni_3Al, *Scripta Mater.* **34** (1996) 4: 527–36

Horvay, M. Theoretische und experimentelle Untersuchung über den Einfluß des inneren Strömungsfeldes auf die Zerstäubungseigenschaften von Drall-Druckzerstäubungsdüsen, Dissertation, Universität Karlsruhe (1985)

Hsiang, L. P. and Faeth, G. M. Near-limit drop deformation and secondary breakup, *Int. J. Multiphase Flow* **18** (1992) 5: 635–52

Hsiang, L. P. and Faeth, G. M. Drop properties after secondary breakup, *Int. J. Multiphase Flow* **19** (1993) 5: 721–35

Hu, H. M., Lavernia, E. J., Lee, Z. H. and White, D. R. Residual stresses in spray-formed A2 tool steel, *J. Mater. Res.* **14** (1999) 12: 4521–4530

Hummert, K. Sprühkompaktieren von Aluminiumwerkstoffen im industriellen Maßstab – Stand der Entwicklung, *Kolloquium des SFB 372*, Vol. 1, Universität Bremen (1996) pp 199–215

Hummert, K. PM-Hochleistungsaluminium im industriellen Maßstab, *Kolloquium des SFB 372*, Vol. 4, Universität Bremen (1999) pp 21–44

Inada, S. and Yang, W. Solidification of molton metal droplets impinging on a cold surface, *Exp. Heat Transfer* **7** (1994) 2: 93–100

Jeffreys, H. On the formation of water waves by wind, *Proc. Roy. Soc. A*, (1924): 189

Jordan, N. und Harig, H. Sprühkompaktierte Kupferbasis-Werkstoffe – Stand der Forschungs- und Entwicklungsarbeiten, *Kolloquium des SFB 372*, Vol. 3, Universität Bremen (1998) pp 31–52

Jordan, N., Schröder, R., Harig, H. and Kienzler, R. Influences of the spray deposition process on the properties of copper and copper alloys, *Mater. Sci. Engng.* **A326** (2002) 1: 51–62

Kallien, L. Herstellung schnell erstarrter und hochunterkühlter Metallpulver, PhD thesis, RWTH, Aachen (1988)

Karl, A. Untersuchung der Wechselwirkung von Tropfen mit Wänden oberhalb der Leidenfrost-Temperatur, PhD thesis, Universität Stuttgart (1997)

Karl, A., Rieber, M., Schelkle, M., Anders, K. and Frohn, A. Comparison of new numerical results for droplet wall interactions with experimental results, *Fluids Engng Division* **236** (1996): 201–6

Kelkar, K. M., Hou, Z., Patankar, S. V., Minisandram, R. S., Forbes Jones, R. M., Carter Jr., W. T., Srivatsa, S. K. and Madden, C. Mathematical model of the clean metal spray forming process, *Proc. 4th International Conference on Spray Forming*, Baltimore MD (1999)

Kienzler, R. and Schröder, R. Entwicklung von Materialmodellen zur Beschreibung des Spannungszustandes und der Porendichte in sprühkompaktierten Komponenten, Sprühkompaktieren, Arbeits- und Ergebnisbericht 1994–1997, *Kolloquium des SFB 372*, Universität Bremen (1997) pp 389–428

Klar, E. and Fesko, J. W. *Powder Metallurgy Metals Handbook*, Vol. 7, American Society for Metals, Materials Park, OH (1984)

Klein, M., Sadiki, A. and Janicka, J. Influence of the inflow conditions on the direct numerical simulation of primary breakup of liquid jets, *ILASS-Europe 2001*, Zürich (2001) pp 475–80

Klein, M., Sadiki, A. and Janicka, J. Untersuchung des Primärzerfalls eines Flüssigkeitsfilms: Vergleich direkte numerische Simulation, Experiment und lineare Theorie, *Spray 2002*, Freiberger Forschungshefte A 870 Verfahrenstechnik, TU-Bergakademie Freiberg (2002) pp 63–72

Knight, R., Smith, R. W. and Lawley, A. Spray forming research at Drexel University, *Int. J. Powder Metall.* **31** (1995) 3: 205–13

Kohnen, G. Über den Einfluß der Phasenwechselwirkungen bei turbulenten Zweiphasenströmungen und deren numerische Erfassung in der Euler-Lagrange Betrachtungsweise, Dissertation, Universität Halle-Wittenberg (1997)

Kothe, D. B. and Mjolsness, R. C. RIPPLE: a new model for incompressible flows with free surfaces, *AIAA Journal* **30** (1992): 11

Kozarek, R. L., León, D. D. and Mansour, A. An investigation of linear nozzles for spray forming aluminium sheets, *Kolloquium des SFB 372*, Vol. 1, Universität Bremen (1996) pp 141–60

Kozarek, R. L., Chu, M. G. and Pien, S. J. An approach to minimize porosity in spray formed deposits through a model-based designed experiment, in: Marsh, S. P. *et al.* (eds.) *Solidification 1998*, The Minerals, Metals & Materials Society, Warrendale, PA (1998) pp 461–71

Kramer, C., Uhlenwinkel, V. and Bauckhage, K. The sticking efficiency at the spray forming of metals, in: Wood, J. V. (ed.) *Proc. 3rd International Conference on Spray Forming*, Cardiff, 1996, Osprey Metals Ltd, Neath (1997)

Kramer, C. Die Kompaktierungsrate beim Sprühkompaktieren von Gauß-förmigen Deposits, Dissertation, Universität Bremen (1997)

Krauss, M., Bergmann, D. and Fritsching, U. In-situ particle temperature, velocity and size measurements in the spray forming process, *Proc. Spray Deposition and Melt Atomization SDMA 2000*, 26–28 June, Bremen, 26–28 June (2000) pp 659–70. Also: *Mater. Sci. Engng.* **A326** (2002) 1: 154–64

Lafaurie, B., Mantel, T. and Zaleski, S. Direct Navier–Stokes simulations of the near-nozzle region, in: Yule, A. J. (ed.) *Proc. ILASS-Europe'98*, 6–8 July, Manchester (1998) pp 54–9

Lafaurie, B., Nardone, C., Scardovelli, R., Zaleski, S. and Zanetti, G. Modelling merging and fragmentation in multiphase flows with SURFER, *J. Comp. Phys.* **133** (1994): 134–47

Lampe, K. Experimentelle Untersuchung und Modellierung der Mehrphasenströmung im düsennahen Bereich einer Öl-Brenner-Düse, Dissertation, Universität Bremen (1994)

Launder, B. E. and Spalding, D. B. The numerical computation of turbulent flows, *Comp. Meth. Appl. Mech. Engng.* **3** (1974): 269–89

Lavernia, E. J. Spray atomization and deposition of metal matrix composites, *Kolloquium des SFB 372*, Vol. 1, Universität Bremen (1996) pp 63–122

Lavernia, E. J., Ayers, J. D. and Srivastan, T. S. Rapid solidification processing with specific application to aluminium alloys, *Int. Mater. Rev.* **37** (1992): 1–44

Lavernia, E. J., Baram, J. and Gutierrez, E. M. Precipitation and excess solid solubility in Mg–Al–Zr and Mg–Zn–Zr alloys processed by spray atomization and deposition, *Mater. Sci. Engng.* **A132** (1991): 119–33

Lavernia, E. J., Gomez, E. and Grant, N. J. The structures and properties of Mg–Zn–Zr and Mg–Zn–Zr alloys produced by LDC, *Mater. Sci. Engng.* **A95** (1987): 225–36

Lavernia, E. J., Rai, G. and Grant, N. J. Rapid solidification processing of 7XXX aluminum alloys: a review, *Mater. Sci. Engng.* **A79** (1986): 211–21

Lavernia, E. J., Gutierrez, E. M., Szekely, J. and Grant, N. J. A mathematical model of the liquid dynamic compaction process. part 1: heat flow in gas atomization, *Int. J. Rapid Solidification* **4** (1988): 89–124

Lavernia, E. J. and Wu, Y. *Spray Atomization and Deposition*, J. Wiley & Sons, Chichester (1996)

Lawley, A. *Atomization – The Production of Metal Powders*, Metal Powder Industries Federation, Princeton, NJ (1992)

Lawley, A. Melt atomization and spray deposition – quo vadis, *Proc. Spray Deposition and Melt Atomization SDMA 2000*, Bremen (2000) pp 3–16

Lawley, A., Mathur, P., Apelian, D. and Meystel, A. Sprayforming: process fundamentals and control, *Powder Metall.* **33** (1990): 109–11

Leatham, A. Spray forming: alloys, products, and markets, *J. Metals* **51** (1999): 4, web-edition: http://www.tms.org/pubs/journals/JOM/9904/Leatham/Leatham-9904.html

Leatham, A. G., Brooks, R. G., Coombs, J. S. and Ogilvy, G. W. in: Wood, J. *Proc. 1st International Conference on Spray Forming*, 17–19 September 1990, Osprey Metals Ltd, Neath, Paper 1 (1991)

Leatham, A. G. and Lawley, A. The Osprey process: principles and applications, *Int. J. Powder Metall.* **29** (1993) 4: 321–9

Lee, E. and Ahn, S. Solidification progress and heat transfer analysis of gas atomized alloy droplets during spray forming, *Acta Metall. Mater.* **42** (1994) 9: 3231–43

Lee, J., Yung, J. Y., Lee, E.-S., Park, W. J., Ahn, S. and Kim, N. J. Dispersion strengthened Cu alloys fabricated in-situ by spray forming, *Kolloquium des SFB 372*, Vol. 4, Universität Bremen (1999) pp 7–20

Lefebvre, A. H. *Atomization and Sprays*, Hemisphere, New York (1989)

Leschziner, M. A. and Rodi, W. Calculation of annular and twin parallel jets using various discretization schemes and turbulence-model variations, *J. Fluids Engng.* **103** (1981): 352–60

Levi, C. G. The evolution of microcrystalline structures in supercooled metal powders, *Metall. Trans. A* **19A** (1988): 699–708

Levi, C. G. and Mehrabian, R. Heat flow during rapid solidification of undercooled metal droplets, *Metall. Trans. A: Phys. Metall. Mater. Sci.* **13A** (1982): 221–34

Levich, V. G. *Physicochemical Hydrodynamics*, Prentice Hall, NJ (1962)

Li, B., Liang, XO., Earthman, J. C. and Lavernia, E. J. Two dimensional modeling of momentum and thermal behaviour during spray atomization of γ-TiAl, *Acta Mater.* **44** (1996) 6: 2409–20

Li, J. PhD thesis, University of Paris VI (1996)

Li, X. Mechanism of atomization of a liquid jet, *Atomiz. & Sprays* **5** (1995): 89–105

Li, X. and Tankin, R. S. Droplet size distribution: a derivation of Nukyama–Tanasawa type distribution function, *Combust. Sci. Technol.* **56** (1987): 65

Liang, X., Earthman, J. C. and Lavernia, E. J. On the mechanism of grain formation during spray atomization and deposition, *Acta Metall. Mater.* **40** (1992) 11: 3003–16

Liang, X. and Lavernia, E. J. Solidification and microstructure evolution during spray atomization and deposition of Ni$_3$Al, *Mater. Sci. Engng.* **A161** (1993): 221–35

Liang, X. and Lavernia, E. J. Evolution of interaction domain microstructure during spray deposition, *Metall. Mater. Trans. A* **25A** (1994): 2341–9

Libera, M., Olsen, G. B. and van der Sande, J. B. Heterogeneous nucleation of solidification in atomized liquid metal droplets, *Mater. Sci. Engng.*, **A132** (1991): 107–18

Liu, H. Berechnungsmodelle für die Geschwindigkeiten und die Abkühlung von Tropfen im Sprühkegel einer Stahl-Zerstäubungsanlage, Dissertation, Universität Bremen (1990)

Liu, H. Numerical modelling of gas atomization in spray forming process, *Proc. 1997 TMS Annual Meeting*, 9–13 February, Orlando, FL (1997)

Liu, H. *Science and Engineering of Droplets: Fundamentals and Applications*, William Andrew, Norwich, NA (2000a)

Liu, H. Spray forming, in: Yu, K. O. *Modelling and Simulation for Casting and Solidification: Theory and Applications*, Marcel Dekker Inc, NY (2000b)

Liu, H., Lavernia, E. J. and Rangel, R. H. Numerical simulation of substrate impact and freezing of droplets in plasma spray processes, *J. Phys. D: Appl. Phys.* **26** (1993): 1900–8

Liu, H., Lavernia, E. J. and Rangel, R. H. Numerical investigation of micropore formation during substrate impact of molten droplets in plasma spray processes, *Atomiz. & Sprays* **4** (1994a): 369–84

Liu. H., Rangel, R. H. and Lavernia, E. J. Modeling of reactive atomization and deposition processing of Ni$_3$Al, *Acta Metall. Mater.* **42** (1994b) 10, 3277–89

Löffler-Mang, M. Düsenströmung, Tropfenentstehung und Tropfenausbreitung bei rücklaufgeregelten Drall-Druckzerstäubern, Dissertation, Universität Karlsruhe (1992)

Love, E., Grisby, C. E., Lee, P. L. and Woodling, M. J. *Experimental and Theoretical Studies of Axisymmetric Free Jets*, NACA Technical Report R-6, Hanover, MD (1959)

Low, T. B. and List, R. Collision, coalescence and breakup of raindrops, *J. Atmosph. Sci.* **39** (1982): 1591–618

Lozano, A., Call, C. J. and Dopazo, C. An experimental and numerical study of the atomization of a planar liquid sheet, *Proc. International Conference on Liquid Atomization and Spray Systems ICLASS '94*, July, Rouen (1994)

Lubanska, H. Correlation of spray ring data for gas atomization of liquid droplets, *J. Metals* **2** (1970): 45–9

Madejski, J. Solidification of droplets on a cold surface, *Int. J. Heat Mass Transfer* **19** (1976): 1009–18

Madejski, J. Droplets on impact with a solid surface, *Int. J. Heat Mass Transfer* **26** (1983): 1095–8

Majagi, S. I., Ranganathan, K., Lawley, A. and Apelian, D. *Microstructural Design by Solidification Processing*, TMS Conference Proceedings, Warrendale, PA (1992) 139 ff.

Malin, M. R. *On the Prediction of Radially Spreading Turbulent Jets*, CHAM Technical Report TR 143, London (1987)

Malot, H. and Dumouchel, C. Volume-based spray drop size distribution: derivation of a generalized gamma distribution from the application of the maximum entropy formalism, CD-ROM *Proc. ILASS-Europe'99*, 5–7 July, Toulouse (1999)

Manson-Whitton, E. D., Stone, I. C., Jones, J. R., Grant, P. S. and Cantor, B. Isothermal grain coarsening of spray formed alloys in the semi-solid state, *Acta Materialia* **50** (2002): 2517–25

Markus, S. and Fritsching, U. Spray forming with multiple atomization, *Proc. Spray Deposition and Melt Atomization SDMA 2003*, 22–25 June, Bremen (2003)

Markus, S., Fritsching, U. and Bauckhage, K. Jet break up of liquid metals, *Proc. Spray Deposition and Melt Atomization SDMA 2000*, 26–28 June, Bremen (2000) pp 497–510. Also: *Mater. Sci. Engng.* **A326** (2002) 1: 122–33

Masuda, W. and Moriyama, E. Aerodynamic characteristics of coaxial impinging jets, *JSME Int. J. Series B*, **37** (1994) 4: 749–75

Mathur, P., Annavarapu, S., Apelian, D. and Lawley, A. Process control, modeling and applications of spray casting, *J. Metals* **41** (1989b): 23–8

Mathur, P., Annavarapu, S., Apelian, D. and Lawley, A. Spray casting: an integral model for process understanding and control, *Mater. Sci. Engng.* **A142** (1991): 261–70

Mathur, P., Apelian, D. and Lawley, A. Analysis of the spray deposition process, *Acta Metall.* **37** (1989a) 2: 429–43

Matteson, M. A., Madden, C. and Moran, A. L. An approach to modelling the spray-forming process with artificial neural networks, *Proc. International Conference on Spray Forming, ICSF-2*, 13–15 September, Swansea (1993)

Maxey, M. R. and Riley, J. J. Equation of motion for a small rigid sphere in a nonuniform flow, *Phys. Fluids* **26** (1983): 883–9

Mayer, W. Zur koaxialen Flüssigkeitszerstäubung im Hinblick auf die Treibstoffaufbereitung in Rake-tentriebwerken, Dissertation, Universität Erlangen (1993)

Medwell, J. O.; Gethin, D. T. and Muhamad, N. Analysis of the Osprey preform deposition process, in *Advances in Powder Matallurgy and Particulate Materials 1992*, Vol. 1: *Powder Production and Spray Forming*, MPIF, Princeton, NJ, pp 249–71

Megaridis, C. M. Presolidification liquid metal droplet cooling under convective conditions, *Atomiz. & Sprays* **3** (1993) 2: 171–91

Menchaca-Rocha, A., Huidobro, F., Martinez-Davalos, A., Michaelian, K., Perez, A., Rodriguez, V. and Carjan, N. Coalescence and fragmentation of colliding mercury drops, *J. Fluid Mech.* **346** (1997): 291–318

Meyer, O., Fritsching, U. and Bauckhage, K. Numerical investigation of alternative process conditions for influencing the thermal history of spray deposited billets, *Proc. Spray Deposition and Melt Atomization SDMA 2000*, 26–28 June, Bremen (2000) pp 771–88

Meyer, O., Fritsching, U. and Bauckhage, K. Numerical investigation of alternative process conditions for influencing the thermal history of spray deposited billets, *Int. J. Thermal Sci.* **42** (2003): 153–68

Meyer, O., Schneider, A., Uhlenwinkel, V. and Fritsching, U. Convective heat transfer from a billet due to an oblique impinging circular jet within the spray forming process, *Int. J. Thermal Sci.* **42** (2003) 6: S561–9

Middleman, S. *Modeling Axisymmetric Flows, Dynamics of Films, Jets, and Drops*, Academic Press, San Diego, CA (1995)

Miles, J. W. On the generation of surface waves by shear flows, Part 1: *J. Fluid Mech.* **3** (1957): 185–204.

Miles, J. W. On the generation of surface waves by shear flows, Part 2: *J. Fluid Mech.* **6** (1958): 568–82

Miles, J. W. On the generation of surface waves by shear flows, Part 3: *J. Fluid Mech.* **7** (1960): 469–478

Miles, J. W. On the generation of surface waves by shear flows, Part 4: *J. Fluid Mech.* **13** (1961): 433–48

Mingard, K. P., Alexander, P. W., Langride, S. J., Tomlinson, G. A. and Cantor, B. Direct measurement of sprayform temperatures and the effect of liquid fraction on microstructure, *Acta Mater.* **46** (1998) 10: 3511–21

Mingard, K. P., Cantor, B., Palmer, I. G., Hughes, I. R., Alexander, P. W., Willis, T. W. and White, J. Macro-segregation in aluminium alloy spray formed billets, *Acta Mater.* **48** (2000): 2435–49

Minisandram, R. S., Forbes Jones, R. M., Kelkar, K. M., Patankar, S. V. and Carter Jr., W. T. Prediction of thermal history of preforms produced by the clean metal spray forming process, *Proc. Spray Deposition and Melt Atomization SDMA 2000*, Bremen (2000) pp 789–802. Also: *Mater. Sci. Engng.* **A326** (2002) 1: 184–93

Moran, A. L. and White, D. R. Developing intelligent control for spray forming processes, *J. Metals* **42** (1990) 7: 21–4

Müller, F. G., Benz, M. G., Carter Jr., W. T., Forbes, R. M. und Leatham, A. Neues Verfahren zur Herstellung von Pulver, Formteilen oder Halbzeugen aus Titan oder keramikfreien Superlegierungen; *Kolloquium des SFB 372*, Vol. 1, Universität Bremen (1996) pp 169–88

Müller, H. R. Eigenschaften und Einsatzpotential sprühkompaktierter Kupferlegierungen, *Kolloquium des SFB 372*, Vol. 1, Universität Bremen (1996) pp 33–56

Mullis, A. M. and Cochrane, R. F. Grain refinement and the stability of dendrites growing into undercooled pure metals and alloys, *J. Appl. Phys.* **82** (1997): 3783–90

Mundo, C. Zur Sekundärzerstäubung newtonscher Fluide an Oberflächen, Dissertation, Universität Erlangen (1996)

Mundo, C., Sommerfeld, M. and Tropea, C. Droplet–wall collisions: experimental studies of the deformation and breakup process, *Int. J. Multiphase Flow* **21** (1995): 151–73

Muoio, N. G., Crowe, C. T., Bergmann, D. and Fritsching, U. Numerical simulation of the turbulent gas-droplet field in spray forming, *3rd International Symposium on Engineering Turbulence Modelling and Measurements*, 27–29 May, Kreta (1996)

Muoio, N. G., Crowe, C. T., Bergmann, D. and Fritsching, U. Numerical simulation of spray temperature in spray forming process by ceramic powder injection, *ASME IMECE Multiphase Flow and Heat Transfer in Materials Processing*, 17–22 November, Atlanta, GA (1996)

Muoio, N. G., Crowe, C. T., Fritsching, U. and Bergmann, D. Modelling metal droplet sprays in spray forming, *ASME Fluids Engng Division* **223** (1995): 111–15

Muoio, N., Crowe, C. T., Fritsching, U. and Bergmann, D. Effect of thermal coupling on numerical simulations of the spray forming process, *Proc. 2nd International Symposium on Numerical Methods for Multiphase Flows*, 7–11 July, San Diego, CA (1996)

Nasr, G. G., Yule, A. J. and Bendig, L. *Industrial Sprays and Atomization: Design, Analysis and Applications*, Springer-Verlag, Heidelberg (2002)

Nichiporenko, O. S. and Naida, Y. I. *Soviet Powder Metallurgy Metal Ceramics* **67** (1968): 509

Nichols, B. D., Hirt, C. W. and Hotchkiss, R. S. *SOLA-VOF: A Solution Algorithm for Transient Fluid Flow with Multiple Free Boundaries*, Report LA-8355, Los Alamos Scientific Laboratory, NM (1980)

Nigmatulin, R. I. *Dynamics of Multiphase Media*, Vols. 1 and 2, Hemisphere, Washington, DC (1990, 1991)

Nobari, M. R. H. and Tryggvason, G. Numerical simulations of three-dimensional drop collisions, *AIAA Journal* **34** (1996): 750–5

Nobari, M. R. H., Jan, Y.-J. and Tryggvason, G. Head-on collision of drops – a numerical investigation, *Phys. Fluids* **8** (1996): 29–42

Norman, A. F., Eckler, K., Zambon, A., Gartner, F., Moir, S. A., Ramous, E., Herlach, D. M. and Greer, A. L. Application of microstructure-selection maps to droplet solidification: a case study of the Ni–Cu system, *Acta Mater.* **46** (1998) 10: 3355–70

Nunez, L. A., Lobel, T. and Palma, R. Atomizers for molten metals: macroscopic phenomena and engineering aspects, *Atomiz. & Sprays* **9** (1999) 6: 581–600

Obermeier, F. (ed.) 7. Workshop über Techniken der Fluidzerstäubung und Untersuchungen von Sprühvorgängen, *Spray 2002*, Freiberger Forschungshefte A 870 Verfahrenstechnik, TU-Bergakademie, Freiberg (2002)

Oertel, H. und Laurien, E. *Numerische Strömungsmechanik*, 2. Aufl. Vieweg Braunschweig (2003)

Ojha, S. N., Jha, J. N. and Singh, S. N. Microstructural modification in Al–Si eutectic alloy produced by spray deposition, *Scripta Metall. Mater.* **25** (1991): 443–7

Ojha, S. N., Tripathi, A. K. and Singh, S. N. Spray atomization and deposition of an Al–4Cu–20Pb alloy, *Powder Metall. Int.* **25** (1993) 2: 65–9

Orme, M. A novel technique of rapid solidification net-form materials synthesis, *J. Mater. Engng. Perform.* **2** (1993) 3: 399–405

Orme, M. and Huang, C. Phase change manipulation for droplet-based solid freeform fabrication, *J. Heat Transfer* **119** (1997): 818–23

Orme, M., Liu, Q. and Fischer, J. Mono-disperse aluminium droplet generation and deposition for net-form manufacturing of structural components, *Proc. International Conference on Liquid Atomization and Spray Systems ICLASS 2000*, 16–20 July, Passadena, CA (2000)

O'Rourke, P. J. Collective drop effects on vaporizing liquid sprays, PhD thesis, Los Alamos National Laboratory, NM (1981)

O'Rourke, P. J. and Amsden, A. A. *The TAB Method for Numerical Calculation of Spray Droplet Breakup*, Report LA-UR-87-2105, Los Alamos National Laboratory, NM (1987)

Ottosen, P. Numerical simulation of spray forming, PhD thesis, Technical University of Denmark, TM.93.27 (1993)

Ozols, A. and Sancho, E. Solidification rates in centrifugal atomisation, *Proc. 1998 PM World Congress*, Vol. 1, 18–22 October, Grenada (1998) pp 179–84

Panchagnula, M. V., Sojky, P. E. and Bajaj, A. K. The non-linear breakup of annular liquid sheets, in: Yule, A. J. (ed.) *Proc. ILASS-Europe'98*, 6–8 July, Manchester, (1998) pp 36–41

Pasandideh-Fard, M., Bhola, R., Chandra, S. and Mostaghimi, J. Deposition of tin droplets on a steel plate: simulations and experiments, *Int. J. Heat Mass Transfer* **41** (1998): 2929–45

Pasandideh-Fard, M., Mostaghimi, J. and Chandra, S. Modeling sequential impact of two molten droplets on a solid surface, *Proc. ILASS-America*, Indianapolis, IN (1999)

Pasandideh-Fard, M., Qiao, Y. M., Chandra, S. and Mostaghimi, J. Capillary effects during droplet impact on a solid surface, *Phys. Fluids* **8** (1996): 650–9

Passow, C. H., Chun, J. H. and Ando, T. Spray deposition of a Sn–40 wt.% Pb alloy with uniform droplets, *Metall. Trans. A* **24A** (1993): 1187–93

Patankar, S. V. *Numerical Heat Transfer and Fluid Flow*, McGraw-Hill, Columbus, OH (1981)

Payne, R. D., Matteson, M. A. and Moran, A. L. Application of neural networks in spray forming technology, *Int. J. Powder Metall.* **29** (1993) 4: 345–51

Payne, R. D., Rebis, A. L. and Moran, A. L. Spray forming quality predictions via neural networks, *J. Mat. Engng. and Perf.* **2** (1996) 5: 693–702

Pedersen, T. P., Hattel, J. H., Proyds, N. H., Pedersen, A. S., Buchholz, M. and Uhlenwinkel, V. A new integrated numerical model for spray atomization and deposition: comparison between numerical results and experiments, *Proc. Spray Deposition and Melt Atomization SDMA 2000*, Bremen (2000) pp 813–24

Pedersen, T. B. Spray forming – a new integrated numerical model, PhD thesis, Technical University of Denmark (2003)

Petersen, K., Pedersen, A. S., Pryds, N., Thorsen, K. A. and List, J. L. The effect of particles in different sizes on the mechanical properties of spray formed steel composites, *Mater. Sci. Engng.* **A326** (2002) 1: 40–50

Pien, S. J., Luo, J., Baker, F. W. and Chyu, M. K. Numerical simulation of a complex spray forming process, *Kolloquium des SFB 372*, Vol. 1, Universität Bremen (1996) pp 161–8

Pilch, M. and Erdmann, C. A. Use of breakup time data and velocity history data to predict the maximum size of stable fragments for acceleration-induced breakup of a liquid drop, *Int. J. Multiphase Flow* **13** (1987): 741–57

Poulikakos, D. and Waldvogel, J. M. Heat transfer and fluid dynamics in the process of spray deposition, *Adv. Heat Transfer* **28** (1996): 1–74

Prakash, C. and Voller, V. On the numerical solution of continuum mixture model equations describing binary solid–liquid phase change, *Numer. Heat Transfer B* **15** (1989): 171–89

Prud'homme, R. and Ordonneau, G. The maximum entropy method applied to liquid jet atomization, CD-ROM *Proc. ILASS-Europe'99*, 5–7 July, Toulouse (1999)

Pryds, N. H. and Hattel, J. H. Numerical modelling of rapid solidification, *Modelling Simul. Mater. Sci. Engng.* **5** (1997): 451–72

Pryds, N., Hattel, J. H., Pedersen, T. B. and Thorborg, J. An integrated numerical model of the spray forming process, *Acta Mater.* **50** (2002): 4075–91

Pryds, N., Hattel, J. H. and Thorborg, J. A quasi-stationary numerical model of atomized metal droplets, II: prediction and assessment, *Modelling Simul. Mater. Sci. Engng.* **7** (1999): 431–46

Quested, P. N., Brooks, R. F., Day, A. P., Richardson, M. J. and Mills, K. C. The physical properties of alloys relevant to spray forming, in: Wood, J. V. (ed.) *Proc. 3rd International Conference on Spray Forming*, 1996, Cardiff, Osprey Metals Ltd, Neath (1997)

Qian, J. and Law, C. K. Regimes of coalescence and separation in droplet collision, *J. Fluid Mech.* **331** (1997): 59–80

Rai, G., Lavernia, E. J. and Grant, N. J. Factors influencing the powder size and distribution in ultrasonic gas atomization, *J. Metals* **37** (1985) 8: 22–6

Rampant Release 4.0.14, Copyright 1996 Fluent Inc. Hanover, NH

Rangel, R. H. and Sirignano, W. A. The linear and nonlinear shear stability of a fluid sheet, *Phys. Fluids* **A3** (1991) 10: 2392–400

Ranz, W. E. and Marshall, W. R. Evaporation from drops – I and II, *Chem. Eng. Prog.* **48** (1952): 141 and 173

Rao, K. P. and Mehrotra, S. P. in: Hausner, H. *et al.* (eds.) *Modern Developments in Powder Metallurgy*, Vol. 12, Metal Powder Industries Federation, Princeton, NJ (1980) pp 113–30

Rau, S. Überprüfung der Eignung von CFD-Simulationsrechnungen zur Ermittlung von Wärmeübergangskoeffizienten auf einer Bolzenoberfläche in einer einphasigen Freistrahlströmung, Studienarbeit thesis, University Bremen (2002)

Rayleigh, Lord, On the stability of jets, *Proc. London Math. Soc.* **10** (1878): 4–13

Rebis, R., Madden, C., Zappia, T. and Cai, C. Computer aided process planning and simulation for the Osprey spray forming process, in: Wood, J. V. (ed.) *Proc. 3rd International Conference on Spray Forming*, 1996, Cardiff, Osprey Metals Ltd, Neath (1997)

Reeks, M. W. and McKee, S. The dispersive effect of Basset history forces on particle motion in a turbulent flow, *Phys. Fluids* **27** (1984) 7: 1573 ff.

Reich, W. and Rathjen, K. D. Numerische Simulation des Tropfenpralls auf feste, ebene Flächen, Studienarbeit Fachgebiet Verfahrenstechnik der Universität Bremen (1990)

Reichelt, W. Stand der industriellen Anwendung des Sprühkompaktierens, *Kolloquium des SFB 372*, Vol. 1, Universität Bremen (1996) pp 189–98

Reichelt, L., Pawlowski, A. and Renz, U. Numerische Untersuchungen zum aerodynamischen Tropfenzerfall mit der Volume-of-Fluid (VOF)-Methode, Freiberger Forschungshefte A 870 Verfahrenstechnik, TU-Bergakademie, Freiberg (2002) pp 133–42

Rein, M. Phenomena of liquid drop impact on solid and liquid surfaces, *Fluid Dynamics Research* **12** (1993): 61–93

Rein, M. The transitional regime between coalescing and splashing drops, *J. Fluid Mech.* 306 (1996): 145–65

Rein, M. Spray deposition: the importance of droplet impact phenomena, *Kolloquium des SFB 372*, Vol. 3, Universität Bremen (1998) pp 115–38

Reitz, R. D. Mechanisms of breakup of round liquid jets, PhD thesis, Princeton University, NJ (1978)

Reitz, R. D. and Bracco, F. V. Mechanism of atomization of a liquid jet, *Phys. Fluids* **25** (1982) 10: 1730–42

Reitz, R. D. and Diwarkar, R. Structure of high pressure fuel sprays, The Engineering Society for Advancing Mobility, Land, Sea, Air, and Space, Warrendale, PA, Paper 870598 (1987)

Ridder, S. D. and Biancaniello, F. S. Process control during high pressure atomization, *Mater. Sci. Engng.* **98** (1988): 47–51

Ridder, S. D., Osella, S. A., Espina, P. I. and Biancaniello, F. S. Intelligent control of particle size distribution during gas atomization, *Int. J. Powder Metall.* **28** (1992) 2: 133–8

Rieber, M. and Frohn, A. Numerical simulation of splashing drops, *Proc. ILASS'98*, 6–8 July, Manchester (1998)

Rioboo, R., Marengo, M. and Tropea, C. Time evolution of liquid drop impact onto solid, dry surfaces, *Exp. Fluids* **33** (2002): 112–24

Roach, S. J., Henein, H. and Owens, D. C. A new technique to measure dynamically the surface tension, viscosity and density of molten metals, *Light Metals* **4** (2001): 1285–91

Roe, P. L. Characteristic based schemes for the Euler equations, *Ann. Rev. Fluid Mech.* **18** (1986): 337–86

Roisman, I. V., Rioboo, R. and Tropea, C. Model for single drop impact on dry surfaces, *Proc. International Conference on Liquid Atomization and Spray Systems ICLASS 2000*, Pasadena, CA (2000)

Roisman, I. V. and Tropea, C. Impact of a drop onto a wetted wall: description of crown formation and propagation, *J. Fluid Mech* **472** (2002): 373–97

Rosten, H. I. and Spalding, D. B. The PHOENICS reference manual, CHAM Technical Report TR/200, London (1987)

Rückert, F. und Stöcker, P. Die neue Alunimium-Silizium-Zylinderlaufbahn-Technologie für Kurbelgehäuse aus Aluminiumdruckguß, *Kolloquium des SFB 372*, Vol. 4, Universität Bremen (1999) pp 45–60

Rüger, M., Hohmann, S., Sommerfeld, M. and Kohnen, G. Euler/Lagrange calculations of turbulent spray: the effect of droplet collisions and coalescence, *Atomiz. & Sprays* **10** (2000) 1: 47–82

Rumberg, O. and Rogg, B. Spray modelling via a joint-PDF formulation for two-phase flow, CD-ROM *Proc. ILASS-Europe'99*, 5–7 July, Toulouse (1999)

Sadhal, S. S., Ayyaswamy, P. S. and Chung, J. N. *Transport Phenomena with Drops and Bubbles*, Mechanical Engineering Series, Springer Verlag, New York (1997)

Samenfink, W., Elsäßer, A. and Dullenkopf, K. Secondary breakup of liquid droplets: experimental investigation for a numerical description, *Proc. Sixth International Conference on Liquid Atomization and Spray Systems ICLASS '94*, 18–22 July, Palais des Congrès, Rouen (1994) pp 156–63

Sanmarchi, C., Liu, H., Lavernia, E. J., Rangel, R. H., Sickinger, A. and Mühlberger, E., Numerical analysis of the deformation and solidification of a single droplet impinging on a flat surface, *J. Mater. Sci.* **28** (1993): 3313–21

Sarkar, S. and Balakrishnan, L. Application of a Reynolds-stress turbulence model to the compressible shear layer, ICASE Report 90–18, NASA CR 182002 (1990)

Scardovelli, R. and Zaleski, S. Direct numerical simulation of free-surface and interfacial flow, *Ann. Rev. Fluid Mech.* **31** (1999): 567–603

Schatz, A. Prozeßsimulation für die Herstellung von Metallschmelzen durch die Gas- und Wasserzerstäubung, lecture skript held at IWT Bremen, 29 September (1994)

Schelkle, M., Rieber, M., and Frohn, A. Comparison of lattice Boltzmann and Navier–Stokes simulations of three-dimensional free surface flows, *Fluids Engng Division* **236** (1996) Fluids Engng. Div. Conf. Vol. 1 ASME 1996: 207–212

Scheller, B. L. and Bousfield, D. W. Newtonian drop impact with a solid surface, *AIChE Journal* **41** (1995) 6: 1357–67

Schmaltz, K. and Amon, C. Experimental verification of an impinging molten metal droplet numerical simulation, *Proc. ASME Heat Transfer Div.* **317** (1995): 219–26

Schmehl, R. CFD analysis of fuel atomization, secondary droplet breakup and spray dispersion in the premix duct of a LPP combustor, 8. *International Conference on Liquid Atomization and Spray Systems ICLASS*, July Passadena, CA, (2000)

Schneider, A., Meyer, O., Tillwick, F., Uhlenwinkel, V. und Fritsching, U. Konvektiver Wärmeübergang an einem schräg angeströmten Bolzen in einer turbulenten Düsenströmung, *Kolloquium des SFB 372*, Vol. 5, Universität Bremen (2001) pp 155–78

Schneider, A. Uhlenwinkel, V. und Bauckhage, K. Zum Ausfließen von Metallschmelzen, *Kolloquium des SFB 372*, Vol. 5, Universität Bremen (2001) pp 69–96

Schneider, S. and Walzel, P. Zerfall von Flüssigkeiten bei Dehnung im Schwerefeld, in: Walzel, P. and Schmidt, D. (eds.) *Proc. SPRAY '98*, 13–14 October, Essen (1998)

Schönung, B. E. *Numerische Strömungsmechanik*, Springer Verlag, Berlin 1990

Schröder, R. und Kienzler, R. Kontinuumsmechanische Untersuchungen an sprühkompaktierten Deposits, *Härt.-Tech.-Mitteilung* **3** (1998a): 172–8

Schröder, R. und Kienzler, R. Numerische Untersuchungen an sprühkompaktierten bolzenförmigen Deposits, *Kolloquium des SFB 372*, Vol. 3, Universität Bremen (1998b) pp 93–114

Schröder, T. *Tropfenbildung an Gerinneströmungen im Schwere- und Zentrifugalfeld*, Fortschr. Ber. VDI Reihe 3: Verfahrenstechnik, No. 503, VDI-Verlag, Düsseldorf (1997)

Schulz, G. Economic production of fine, prealloyed MIM powders by the NANOVAL gas atomization process, *Adv. Powder Metall. & Part. Materials – 1996*, Metal Powder Industries Federation, Princeton, NJ (1996) pp 1–35 – 1–41

Sellens, R. W. and Brzustowski, T. A. A prediction of the drop size distribution in a spray from first principles, *Atomiz. & Spray Technol.* **I** (1985): 85

Sellens, R. W. Prediction of the drop size and velocity distribution in a spray based on the maximum entropy formalism, *Part. Part. Syst. Charact.* **6** (1989): 17

Seok, H. K., Lee, H. C., Oh, K. H., Lee, J.-C., Lee, H. I. and Ra, H. Y. Formulation of rod forming models and their application to spray forming, *Metall. Mater. Trans. A* **31A** (2000): 1479–88

Seok, H. K., Yeo, D. H., Oh, K. H., Lee, J.-C., Lee, H.-I. and Ra, H. Y. A three-dimensional model of the spray forming method, *Metall. Mater. Trans. A* **29** (1998): 699–708

Seok, H. K., Yeo, D. H., Oh, K. H., Ra, H. Y. and Shin, D. S. 3-dimensional forming model of rod in spray forming method, in: Wood, J. V. (ed.) *Proc. 3rd International Conference on Spray Forming*, 1996, Cardiff, Osprey Metals Ltd, Neath (1997)

Shan, X. and Chen, H. Simulation of non-ideal gases and liquid–gas phase transitions by the lattice Boltzmann equation, *Phys. Rev. E* **49** (1994): 2941–8

Shannon, C. E. and Weaver, W. *The Mathematical Theory of Communication*, University of Illinois Press, Urbana, IL (1949)

Shokoohi, F. and Elrod, H. G. Numerical investigation of the disintegration of liquid jets, *J. Comput. Phys.* **71** (1987): 324–42

Shukla, P., Mandal, R. K. and Ojha, S. N. Non-equilibrium solidification of undercooled droplets during atomization process, *Bull. Mater. Sci.* **24** (2001) 5: 547–54

Shukla, P., Mishra, N. S. and Ojha, S. N. Modeling of heat flow and solidification during atomization and spray deposition processing, *J. Thermal Spray Technol.* **12** (2003) 1: 95–100

Singer, A. R. E. The principle of spray rolling of metals, *Metal Mater.* **4** (1970): 246–50

Singer, A. R. E. British Patent No. 1, 262, 471 (1972a)

Singer, A. R. E. Aluminium and aluminium-alloy strip produced by spray deposition and rolling, *J. Inst. Metals* **100** (1972b): 185–90

Singer, A. R. E. Recent developments and opportunities in spray forming, *Kolloquium des SFB 372*, Vol. 1, Universität Bremen (1996) pp 123–40

Singh, R. P., Lawley, A., Friedman, S. and Nurty, Y. V. Microstructure and properties of spray cast Cu–Zr alloys, *Mater. Sci. Engng.* **A145** (1991): 243–55

Sirignano, W. A. *Fluid Dynamics and Transport of Droplets and Sprays*, Cambridge University Press, New York (1999)

Sizov, A. M. *Dispersion of Melts by Supersonic Gas Jets*, Metallurgija, Verlag Moskau (1991)

Smith, M. F., Neiser, R. A. and Dykhuizen, R. C. *NTSC'94, Proc. 7th National Thermal Spray Conference*, 12–15 June, American Society of Metals, Boston, MA (1994)

Sommerfeld, M. *Modellierung und numerische Berechnung von partikelbeladenen turbulenten Strömungen mit Hilfe des Euler/Lagrange Verfahrens*, Verlag Shaker, Aachen (1996)

Song, J. L., Dowson, A., Jacobs, M. H., Brooks, J. K. and Beden, I. FE simulation of the ring rolling process and the implications of prior processing by low pressure centrifugal spray deposition, *Adv. Technol. Plasticity* **12** (1999): 2419–24

Spalding, D. B. The calculation of free-convection phenomena in gas–liquid mixtures, ICHMT Seminar, Dubrovnik (1976)

Spalding, D. B. *Combustion and Mass Transfer*, 1st Edn., Pergamon Press, Oxford (1979)

Spiegelhauer, C., Shaw, L., Overgaard, J. und Oaks, G. Horizontale Sprühkompaktierung von großen Bolzen aus Fe-Legierung, *Kolloquium des SFB 372*, Vol. 1, Universität Bremen (1996) pp 57–62

Spiegelhauer, C. Properties of spray formed highly alloyed tool steels, *Kolloquium des SFB 372*, Vol. 3, Universität Bremen (1998) pp 1–8

Spiegelhauer, C. State of the art for making tool steel billets by spray forming, *Kolloquium des SFB 372*, Vol. 5, Universität Bremen (2001) pp 63–8

Srivastava, V. C., Mandal, R. K. and Ojha, S. N. Monte Carlo simulation of droplet mass flux during gas atomization and deposition, *Proc. Spray Deposition and Melt Atomization SDMA 2000*, Bremen (2000) pp 855–868

STAR-CD, User Guide, Computational Dynamics Ltd (1999)

Sterling, A. M. and Schleicher, C. A. The instability of capillary jets, *J. Fluid Mech.* **68** (1975): 477–95

Sterling, T. L., Salmon, J., Becker, D. J. and Savarese, D. F. *How to Build a Beowulf, A Guide to the Implementation and Application of PC Clusters*, MIT Press, Cambridge, MA (1999)

Su, Y. H. and Tsao, C. Y. A. Modeling of solidification of molten metal droplets during atomization, *Metall. Mater. Trans.* **28B** (1997): 1249–55

Taylor, G. I. *Generation of Ripples by Wind Blowing over a Viscous Liquid*, Collected Works of G. I. Taylor, Vol. 3 (1940)

The Aluminum Association, *Aluminum Industry Technology Roadmap*, The Aluminum Association, Washington, DC (1997)

Tillwick, F. Ermittlung von Wärmeübergangskoeffizienten an der Bolzenoberfläche in einer einphasigen Freistrahlströmung, Studienarbeit, Universität Bremen (2000)

Ting, J., Peretti, M. W. and Eisen, W. B. The effect of deep aspiration on gas-atomized powder yield, *Proc. Spray Deposition and Melt Atomization SDMA 2000*, Bremen (2000) pp 483–96, Also: *Mater. Sci. Engng.* **A326** (2002) 1: 110–21

Tinscher, R., Bomas, H. und Mayr, P. Untersuchungen zum Sprühkompaktieren des Stahls 100Cr6, *Kolloquium des SFB 372*, Vol. 4, Universität Bremen (1999) pp 77–104

Torrey, M. D., Cloutman, L. D., Mjolsness, R. C. and Hirt, C. W. *NASA-VOF2D: A Computer Program for Incompressible Flows with Free Surfaces*, Report LA-10612-MS, Los Alamos Scientific, NM (1985)

Torrey, M. D., Mjolsness, R. C. and Stein, L. R. *NASA-VOF3D: A Three-Dimensional Computer Program for Incompressible Flows with Free Surfaces*, Report LA-11009-MS Los Alamos Scientific Laboratory, NM (1987)

Trapaga, G., Matthys, E. F., Valencia, J. J. and Szekely, J. Fluid flow, heat transfer and solidification of molten metal droplets impinging on substrates – comparison of numerical and experimental results, *Metall. Trans. B Process Metall.* **23B** (1992): 701–18

Trapaga, G. and Szekely, J. Mathematical modeling of the isothermal impingement of liquid droplets in spray forming, *Metall. Trans. B* **22** (1991): 901–10

Truckenbrodt, E. *Fluidmechanik, Vol. 1: Grundlagen und elementare Strömungsvorgänge dichtebeständiger Fluide*, Springer-Verlag, Berlin u.a., 3. Aufl. (1989)

Truckenbrodt, E. *Fluidmechanik, Vol. 2: Elementare Strömungsvorgänge dichteveränderlicher Fluide sowie Potential- und Grenzschichtströmungen*, Springer-Verlag, Berlin u.a. (1980)

Tsao, C.-Y. A. and Grant, N. J. Modeling of the liquid dynamic compaction spray process, *Int. J. Powder Metall.* **30** (1994) 3: 323–33

Tsao, C.-Y. A. and Grant, N. J. Microstructure and recrystallization behaviour of in-situ alloyed and microalloyed spray-formed SAE 1008 steel, *Kolloquium des SFB 372*, Vol. 4, Universität Bremen (1999) pp 61–76

Turnbull, D. Formation of crystal nuclei in liquid metals, *J. Appl. Phys.* **21** (1950): 1022–8

Uhlenwinkel, V. Zum Ausbreitungsverhalten der Partikeln bei der Sprühkompaktierung von Metallen, Dissertation, Universität Bremen (1992)

Uhlenwinkel, V., Fritsching, U. and Bauckhage, K. The influence of spray parameters on local mass fluxes and deposit growth rates during spray compaction process, *Proc. 5th International Conference on Liquid Atomization and Spray Systems, ICLASS-91*, Gaithersburg, MD, NIST SP813 (1991) pp 483–90

Uhlenwinkel, V., Fritsching, U., Bauckhage K. und Urlau, U. Strömungsuntersuchungen im Düsennahbereich einer Zweistoffdüse – Modelluntersuchungen für die Zerstäubung von Metallschmelzen, *Chem.-Ing. Tech.* **62** (1990) 3: 228–9, Synopse 1840

Ünal, A. Effect of processing variables on particle size in gas atomization of rapidly solidified aluminium powders, *Mater. Sci. Technol.* **3** (1987): 1029–39

Ünal, A. Flow separation and liquid rundown in a gas-atomization process, *Metall. Trans.* **20B** (1989): 613–22

Underhill, R. P., Grant, P. S., Bryant, D. J. and Cantor, B. Grain growth in spray-formed Ni superalloys, *J. Mater. Synthesis Processing* **3** (1995) 3: 171–9

van der Sande, E. and Smith, J. M. Jet breakup and air entrainment by low-velocity turbulent jets, *Chem. Engng. Sci.* **31** (1973) 3: 219–24

Vardelle, A., Themelis, N. J., Dussoubs, B., Vardelle, M. and Fauchais, P. Transport and chemical rate phenomena in plasma sprays, *High Temp. Chem. Process* **1** (1997) 3: 295–313

Venekateswaren, S., Weiss, J. M. and Merkle, C. L. *Propulsion Related Flowfields Using Preconditioned Navier–Stokes Equations*, Technical Report AAIA-92-3437, American Institute of Aeronautics and Astronautics, Reston, VA (1992)

Voller, V. R., Swaminathan, C. R. and Thomas, B. G. Fixed grid techniques for phase change problems: a review, *Int. J. Num. Methods Engng.* **30** (1990): 875–98

Voller, V. R., Swaminathan, C. R. and Thomas, B. G. General source-based methods for solidification phase change, *Num. Heat Transfer, Part B* **19** (1991): 175–89

Waldvogel, J. M. and Poulikakos, D. Solidification phenomena in picoliter size solder droplet deposition on a composite substrate, *Int. J. Heat Mass Transfer* **40** (1997): 295–309

Walzel, P. Zerteilgrenze beim Tropfenaufprall, *Chem.-Ing.-Tech.* **52** (1980): 338–9

Wang, G. X. and Matthys, E. F. Modelling of heat transfer and solidification during splat cooling: effect of splat thickness and splat/substrate thermal contact, *Int. J. Rapid Solidification* **6** (1991): 141–74

Wang, G. X. and Matthys, E. F. Numerical modelling of phase change and heat transfer during rapid solidification processes: use of control volume integral with element subdivision, *Int. J. Heat Mass Transfer* **35** (1992): 141–53

Weber, C. Zum Zerfall eines Flüssigkeitsstrahles, *Z. Angew. Math. Mech.* **11** (1931): 138–45

Weiss, J. M. and Smith, W. A. Preconditioning applied to variable and constant density flows, *AIAA Journal* **33** (1995): 2050–57

Weiss, D. A. and Yarin, A. L. Single drop impact onto liquid films: neck distortion, jetting, tiny bubble entrainment, and crown formation, *J. Fluid Mech.* **385** (1999): 229–54

Welch, J. E., Harlow, F. H., Shannon, J. P. and Daly, B. J. *The MAC-Method: A Computing Technique for Solving Viscous, Incompressible, Transient Fluid-Flow Problems Involving Free Surfaces*, Report LA-3425, Los Alamos Scientific Laboratory, NM (1966)

Winnikow, S. and Chao, B. T. Droplet motion in purified systems, *Phys. Fluids* **9** (1965) 1: 50–61

Wood, J. V. (ed.) *Proc. 2nd International Conference on Spray Forming*, Swansea, 1993, Woodhead, Cambridge (1993)

Wood, J. V. (ed.) *Proc. 3rd International Conference on Spray Forming*, Cardiff, 1996, Osprey Metals Ltd, Neath (1997)

Wood, J. V. (ed.) *Proc. 4th International Conference on Spray Forming*, Baltimore, MD 1999, Osprey Metals Ltd and Welsh Development Agency, Neath (1999)

Woodruff, D. P. *The Solid–Liquid Interface*, Cambridge University Press (1973)

Wu, Y., Zhang, J. and Lavernia, E. J. Modeling of the incorporation of ceramic particulates in metallic droplets during spray atomization and coinjection, *Metall. Mater. Trans. B* **25** (1994): 135–47

Wünnenberg, K. Sprühkompaktieren von Stahl: Verfahrenstechnik und Produkteigenschaften, *Kolloquium des SFB 372*, Vol. 1, Universität Bremen (1996) pp 1–32

Xu, Q. and Lavernia, E. J. Numerical calculations of heat transfer and nucleation in the initially deposited material during spray atomization and deposition, *Proc. 4th International Conference on Spray Forming ICSF*, 13–15 September, Baltimore, MD (1999)

Xu, Q. and Lavernia, E. J. Fundamentals of the spray forming process, *Proc. Spray Deposition and Melt Atomization SDMA 2000*, Bremen (2000) pp 17–36

Yakhot, V. and Orszag, S. A. Renormalization group analysis of turbulence, I. basic theory, *J. Sci. Comput.* **1** (1986): 3–51

Yang, B., Wang, F., Cui, H., Duan, B. Q. and Zhang, J. S. Research and development of spray deposited material in China, *Proc. Spray Deposition and Melt Atomization SDMA 2000*, Bremen (2000) pp 53–60

Yarin, A. L. and Weiss, D. A. Impact of drops on solid surfaces: self-similar capillary waves, and splashing as a new type of kinematic discontinuity, *J. Fluid Mech.* **283** (1995): 141–73

Yearling, P. R. and Gould, R. D. Convective heat and mass transfer from single evaporating water, methanol and ethanol droplets, *ASME Fluids Engng Division* **223** (1995): 33–8

Yule, A. J. and Dunkley, J. J. *Atomization of Melts*, Clarendon Press, Oxford (1994)

Zaleski, S. and Li, J. Direct simulation of spray formation, *Proc. International Conference on Liquid Atomization and Spray Systems ICLASS '97*, Seoul, August (1997) pp 812–19

Zaleski, S., Li, J., Succi, S., Scardovelli, R. and Zanetti, G. Direct numerical simulation of flows with interfaces, *Proc. 2nd International Conference on Multiphase Flow*, April, Kyoto (1995)

Zeng, X. and Lavernia, E. J. Interfacial behaviour during spray atomization and co-deposition, *Int. J. Rapid Solidification* **7** (1992): 219–43

Zhang, H. Temperaturverteilung im aufwachsenden Deposit und im Substrat sowie Verläufe des Erstarrungsgrades im Deposit bei der Sprühkompaktierung von Metallen, Dissertation, Universität Bremen (1994)

Zhang, J., Wu, Y. and Lavernia, E. J. Kinetics of ceramic particulate penetration into spray atomized metallic droplets at variable penetration depth, *Acta Metall. Mater.* **42** (1994): 2955–72

Zhao, Y. Y., Dowson, A. L. and Jacobs, M. H. Modelling of liquid flow after a hydraulic jump on a rotating disc prior to centrifugal atomization, *Modelling Simul. Mater. Sci. Eng.* **8** (2000) 1: 55–65

Zhao, Y. Y., Dowson, A. L., Johnson, T. P., Young, J. M. and Jacobs, M. H. Prediction of liquid metal velocities on a rotating disk in spray forming by centrifugal spray deposition, *Adv. Powder Metall. & Part. Materials – 1996*, Metal Powder Industries Federation, Princeton, NJ (1996a) pp 9–79 – 9–89

Zhao, Z., Poulikakos, D. and Fukai, J. Heat transfer and fluid dynamics during the collision of a liquid droplet on substrate – I. modeling, *Int. J. Heat Mass Transfer* **39** (1996b): 2771–89

Zhou, Z.-W. and Tang, X.-D. The effect of the pulsation in gas flow on the stability of molten metal jet, *Proc. 4th International Conference on Spray Forming ICSF*, Baltimore, MD (1999)

Useful web pages

Author's homepage

http://www.iwt-bremen.de/vt/MPS/

Modelling and simulation – CFD

http://www.cfd-online.com/
http://www.mie.utoronto.ca/labs/tsl/
http://capella.colorado.edu/~laney/software.htm

Atomization and sprays

http://www.atomization.de/
http://www.me.umist.ac.uk/asrgpage/index.htm
http://www.ilass-eu.ic.ac.uk/ilass_eu.htm
http://www.ilass.uci.edu/
http://www.atomising.co.uk/

Spray forming

http://sfb372.iwt.uni-bremen.de/home-a.html
http://gram.eng.uci.edu/~sprays/
http://www.irc.bham.ac.uk/theme1/
http://users.ox.ac.uk/~pgrant/sfintro.html
http://www.arl.psu.edu/areas/spraymetform/spraymetform.html
http://www.ospreymetals.co.uk/
http://www.dansteel.dk/Danspray/Danspray/html/DanSpray_uk.htm

Index

algebraic solution, 25
atomization, 26
averaging method, 138

back face culling, 184
back-splashing, 43
bag break-up, 90
balance force, 98
Basset–Boussinesq–Oseen, 98
Basset history integral, 99
batch process, 33
billet cooling, 187
billet surface model, 216
Boltzmann equation, 38
boundary conditions, 29
boundary fitted, 24

caloric averaging, 137
centrifugal spray, 35
clean spray forming process, 216
close-coupled, 80
close-coupled configuration, 27
coaxial gas flow field, 42
collision probabilities, 150
compacting process, 161
compaction, 6
conjugate heat transfer, 30
conservation equations, 21
conservation, 22
continuum surface, 37
control, 17
cooling control, 213
coordinate transformation, 188
crucible, 6

dilute flow, 149
direct numerical simulation, 85, 96
discrete vortex methods, 96
disintegration, 6, 12
disintegration efficiency, 52
dissipation rate, 22
division of spray forming, 10
(drag) coefficient, 100
droplet and particle, 147

droplet impact, 161
droplet solidification, 169

Euler/Euler, 37
Eulerian/Eulerian, 97
Eulerian/Lagrangian, 96

form-filling spray process, 180
fragmentation delay, 89
free-fall atomizer, 27
freezing, 27
full billet model, 216

geometric modelling, 176
geometries of spray forming, 8
grain-size modelling, 225

heat resistance coefficient, 192
heat transfer, 107
high-resolution models, 14
homogeneous two-phase model, 37
hot workability, 9

ideal expanded gas, 61
independent process conditions, 14
industrial applications, 7
integral, 17
integral modelling, 233
interacting gas jet systems, 55
intermetallic composite, 8
inviscid theory, 29

Kelvin–Helmholtz, 60

large eddy simulation, 96
large-wavelength, 69
laser Doppler anemometry, 43
Laval nozzles, 62
low-resolution models, 14
Lubanska's formula, 78

macro- and micopore formation, 219
macropores, 177
macro segregations, 9
main modes, 89

marker and cell, 36
mass yield, 14
material properties, 13
maximum entropy formalism, 83
melt delivery, 10
melt disintegration, 6
metal casting, 6
metal-matrix-composites, 8, 156
metallurgical aspects, 13
microstructure, 9
microstructure modelling, 222
modified dissipation, 56
multicoupled simulation, 95
multimode, 90
multipurpose, 24
mushy layer, 209

near-net shape, 6
neural, 18
non-linear stability analyses, 71
numerical model, 23

operational models, 1
operational parameters, 131
Osprey process, 7
overexpanded jets, 61
oxygen contamination, 9

partial fragmentation, 163
particle/particle interactions, 13
particle penetration mechanism, 156
particle size spectrum, 18
particle-source-in-cell, 122
phase-Doppler-anemometry, 101
physical model, 21
piecewise linear, 85
planning models, 1
porosity factor, 177
powder metallurgy, 6
primary liquid, 75
probability density functions, 96
product, 13

Ranz and Marshall, 108
rebounding, 163
recalescence, 112
residual stress, 219

secondary fragmentation, 81
sensors, 18
small-wavelength, 70
solidification modelling, 109
source or sink terms , 21
span of the droplet-size distribution, 63
splashing number, 176
splashing of the droplet, 163
spray, 6
spray analysis, 12
spray structure, 12
spray-chamber, 144
stability analysis, 68
Stokes resistance, 122
subsonic flow, 44
super alloys, 8
supersonic flow, 52
surface temperature, 237
surface tension, 41

the gas pressure, 18
theoretical atomization, 44
thermal averaging, 137
thermal model, 20
tracer-particle, 45
transformation of, 57
turbulence, 22
turbulent dispersion, 105
turbulent droplet, 135
turbulent kinetic energy, 22
twin-fluid atomization, 6
twin-fluid atomizers, 26
two-layer zonal model, 197
two-phase flow, 12

undercooling, 111
underexpanded jets, 53
unit operations, 1

visualization, 25
volume, 36
vortex formation, 29

wall friction, 32

yield strength, 9